Within Reach

Within Reach

Navigating the Political Economy of Decarbonization

Stéphane Hallegatte, Catrina Godinho, Jun Rentschler, Paolo Avner, Ira Irina Dorband, Camilla Knudsen, Jana Lemke, and Penny Mealy

© 2024 International Bank for Reconstruction and Development / The World Bank
1818 H Street NW, Washington, DC 20433
Telephone: 202-473-1000; Internet: www.worldbank.org

Some rights reserved

1 2 3 4 27 26 25 24

This work is a product of the staff of The World Bank with external contributions. The findings, interpretations, and conclusions expressed in this work do not necessarily reflect the views of The World Bank, its Board of Executive Directors, or the governments they represent. The World Bank does not guarantee the accuracy, completeness, or currency of the data included in this work and does not assume responsibility for any errors, omissions, or discrepancies in the information, or liability with respect to the use of or failure to use the information, methods, processes, or conclusions set forth. The boundaries, colors, denominations, and other information shown on any map in this work do not imply any judgment on the part of The World Bank concerning the legal status of any territory or the endorsement or acceptance of such boundaries.

Nothing herein shall constitute or be construed or considered to be a limitation upon or waiver of the privileges and immunities of The World Bank, all of which are specifically reserved.

Rights and Permissions

This work is available under the Creative Commons Attribution 3.0 IGO license (CC BY 3.0 IGO) http://creativecommons.org/licenses/by/3.0/igo. Under the Creative Commons Attribution license, you are free to copy, distribute, transmit, and adapt this work, including for commercial purposes, under the following conditions:

Attribution—Please cite the work as follows: Hallegatte, Stéphane, Catrina Godinho, Jun Rentschler, Paolo Avner, Ira Irina Dorband, Camilla Knudsen, Jana Lemke, and Penny Mealy. 2024. *Within Reach: Navigating the Political Economy of Decarbonization*. Climate Change and Development Series. Washington, DC: World Bank. doi:10.1596/978-1-4648-1953-7. License: Creative Commons Attribution CC BY 3.0 IGO

Translations—If you create a translation of this work, please add the following disclaimer along with the attribution: This translation was not created by The World Bank and should not be considered an official World Bank translation. The World Bank shall not be liable for any content or error in this translation.

Adaptations—If you create an adaptation of this work, please add the following disclaimer along with the attribution: This is an adaptation of an original work by The World Bank. Views and opinions expressed in the adaptation are the sole responsibility of the author or authors of the adaptation and are not endorsed by The World Bank.

Third-party content—The World Bank does not necessarily own each component of the content contained within the work. The World Bank therefore does not warrant that the use of any third-party-owned individual component or part contained in the work will not infringe on the rights of those third parties. The risk of claims resulting from such infringement rests solely with you. If you wish to reuse a component of the work, it is your responsibility to determine whether permission is needed for that reuse and to obtain permission from the copyright owner. Examples of components can include, but are not limited to, tables, figures, or images.

All queries on rights and licenses should be addressed to World Bank Publications, The World Bank, 1818 H Street NW, Washington, DC 20433, USA; e-mail: pubrights@worldbank.org.

ISBN (paper): 978-1-4648-1953-7
ISBN (electronic): 978-1-4648-1954-4
DOI: 10.1596/978-1-4648-1953-7

Cover image: © Arthimedes / Shutterstock. Used with the permission of Arthimedes / Shutterstock. Further permission required for reuse.
Cover design: Bill Pragluski, Critical Stages, LLC

Library of Congress Control Number: 2023919988

Climate Change and Development

The Climate Change and Development Series was created in 2015 to showcase economic and scientific research that explores the interactions between climate change, climate policies, and development. The series aims to promote debate and broaden understanding of current and emerging questions about the climate-development nexus through evidence-based analysis.

The series is sponsored by the Sustainable Development Vice Presidency of the World Bank, and its publications represent the highest quality of research and output in the institution on these issues. The World Bank is committed to sharing relevant and rigorously peer-reviewed insights on the opportunities and challenges present in the climate-development nexus with policy makers, the academic community, and a wider global audience.

TITLES IN THIS SERIES

Green Digital Transformation: How to Sustainably Close the Digital Divide and Harness Digital Tools for Climate Action (2024) by the World Bank

Within Reach: Navigating the Political Economy of Decarbonization (2024) by Stéphane Hallegatte, Catrina Godinho, Jun Rentschler, Paolo Avner, Ira Irina Dorband, Camilla Knudsen, Jana Lemke, and Penny Mealy

Reality Check: Lessons from 25 Policies Advancing a Low-Carbon Future (2023) by the World Bank

Diversification and Cooperation in a Decarbonizing World: Climate Strategies for Fossil Fuel-Dependent Countries (2020) by Gregorz Peszko, Dominique van der Mensbrugghe, Alexander Golub, John Ward, Dimitri Zenghelis, Cor Marijs, Anne Schopp, John A. Rogers, and Amelia Midgley

Unbreakable: Building the Resilience of the Poor in the Face of Natural Disasters (2017) by Stéphane Hallegatte, Adrien Vogt-Schilb, Mook Bangalore, and Julie Rozenberg

Shock Waves: Managing the Impacts of Climate Change on Poverty (2016) by Stéphane Hallegatte, Mook Bangalore, Laura Bonzanigo, Marianne Fay, Tamaro Kane, Ulf Narloch, Julie Rozenberg, David Treguer, and Adrien Vogt-Schilb

Decarbonizing Development: Three Steps to a Zero-Carbon Future (2015) by Marianne Fay, Stéphane Hallegatte, Adrien Vogt-Schilb, Julie Rozenberg, Ulf Narloch, and Tom Kerr

Contents

Boxes

Figures

Maps

Tables

Foreword

Political economy can be a sensitive topic. At an individual level, why people do what they do and think what they think is inherently personal, private even. Collectively, however, people's thoughts, feelings, and behaviors shape current and future events.

It is tempting to avoid analyzing or addressing the political economy for fear of creating unnecessary controversy. But for climate policy makers, this stance will not hold, as shown by the successes and failures of the past decade.

The world has united around the Paris Agreement on climate change, committing to hold global temperature rise to less than 2°C above preindustrial levels and pursue efforts to limit it to 1.5°C. More than 70 countries, representing 76 percent of global emissions, have pledged to reach net zero emissions. Meanwhile, the costs of low-carbon technologies have fallen, and their capabilities have risen.

If people were driven purely by science and economics, the climate crisis would be halfway to solved by now. But, as the latest Intergovernmental Panel on Climate Change (IPCC) report lays out so unequivocally, it is not. Moreover, unless the political economy is considered as thoroughly as the science and economics of climate change, it will not be.

When we look at climate success stories across the world, it is tempting to focus on the substance of policies, be it climate-smart agriculture, renewable energy, or green building codes. This is undoubtedly useful. Yet it is equally useful to peel back the particulars to expose the underlying characteristics of successful policies. In doing so, we are forced to reevaluate what is considered possible.

As this book shows, overcoming political economy barriers is within reach. However, it does require policy makers to adopt certain approaches.

First, appreciate that the political economy is not a static force to navigate. Rather, it is a dynamic relationship that evolves. Indeed, policy makers have the ability to strategically shape how the political economy evolves by doing things that build support over time.

Politicians build support for what they want to do by knowingly picking their battles and words to change minds. Policy makers can also bridge the gap between what is possible and what is needed by knowingly advancing policies that will be well received by most, if not all. This, in turn, facilitates further action. As the adage goes, "new policy creates new politics."

Second, fully consider what is really motivating people. People's willingness to embrace climate policies is not purely driven by fear of extreme climate impacts or whether they themselves personally benefit. There are even cases where direct beneficiaries of particular climate policies still oppose them because they perceive them as unfair or illegitimate.

Sometimes, the primary driver of individuals' views on climate action is not their budget but their beliefs. Yes, policy makers need the hard data to plainly assess where the costs and benefits fall. But it is essential to acknowledge that this is only half the equation.

Third, embrace pragmatism. This is easier said than done. As the science becomes increasingly grim and the timeline more urgent, it is tempting to become more unyielding in what is considered an adequate policy response. Fight that urge. It will only close doors. Instead, be more flexible about how to achieve climate objectives on the informed understanding that doing so will open more doors than it closes.

In practice, that requires policy makers to give much more thought to policy opponents—both the substance behind their opposition as well as the power they wield to slow or sink action. Find out what feeds them, without presumption, and make room for them.

The need for emissions reduction is more urgent than ever. The approach proposed in this book is not about slowing change—quite the opposite. By starting with what is possible, policy makers can create momentum and help catalyze new technologies, new economics, and new politics, making accelerated change possible.

Navigating political economy barriers is hard. It is easier to believe that if people just understand the science of climate change, they will support emissions reduction efforts. Or that if people benefit from a climate policy, they will support it. That may be true to a degree, but it will not ever be enough.

Knowing what has been achieved, and the urgency of what is left to do, policy makers need to favor climate action that is dynamic, that allows for the paradoxes of human nature, and that is above all pragmatic. The success of climate action over the next decade and beyond depends upon policy makers embracing their power to shape not only economic and technological fortunes, but the political economy too.

Juergen Voegele
Vice President for Sustainable Development
World Bank

Acknowledgments

The preparation of the report was led by Stéphane Hallegatte, Catrina Godinho, and Jun Rentschler, with a core team comprising Paolo Avner, Ira Irina Dorband, Camilla Knudsen, Jana Lemke, and Penny Mealy. Stéphane Hallegatte supervised the preparation, review, and editing of the report. Catrina Godinho co-led the writing of the report and project administration, conceptualization, review, and editing. Jun Rentschler co-led the conceptualization, project management, and analysis. Penny Mealy led the preparation of chapter 3; Ira Irina Dorband and Paolo Avner led the preparation of chapter 4; and Camilla Knudsen led the preparation of chapters 5. Jana Lemke was instrumental in writing, as well as review and production. All authors contributed to writing, reviewing, and preparing the final manuscript.

The report includes important contributions and inputs from Alina Averchenkova, Puneet Chitkara, Mathieu Cloutier, Jose Antonio Cuesta, Hancheng Dai, Frederick Daley, Verena Fritz, Michael Ganslmeier, Elisha George, Meghan Gordon, David Groves, Marek Hanusch, Gilang Hardadi, Daniel Herszenhut, Kayenat Kabir, Tamara Krawchenko, Mariza Montes de Oca Leon, Jia Li, Charlotte Liotta, Vivian Liu, Alexander Lotsch, Jorge Martinez-Vazquez, Nicholas Menzies, Tom Moerenhout, Jonas Nahm, Shohei Nakamura, Andrew Nell, Peter Newell, Samuel Okullo, Wei Peng, Joe Pryor, Sophie de Vries Robbe, Marcus Saraiva, Hugh Searight, Harris Selod, Anjali Sharma, Iryna Sikora, Johannes Urpelainen, Vincent Viguié, Anya Vodopyanov, and Farah Zahir.

Key feedback and suggestions were provided by Luc Christiaensen, Joeri de Wit, Joseph Dickman, Chandrasekar Govindarajalu, Dirk Heine, Stuti Khemani, Xenia Vanessa Kirchhofer, Somik Lall, Alan David Lee, Neil McCulloch, Sheoli Pargal, Rachel Bernice Perks, Julie Rozenberg, Marcela Rozo, Neha Sharma, Michael Stanley, Brian James Walsh, Michael Ward, and Melvin H. L. Wong.

We gratefully acknowledge the invaluable comments and advice provided at different stages by internal and external peer reviewers: Aziz Amuzaini, Eric Arias, Christian Bodewig, Chiara Bronchi, Elizabeth N. Ruppert Bulmer, Kevin Carey, Louise J. Cord, Thomas Farole, Marianne Fay, Caroline Fischer, Vivien Foster, Adrian Fozzard, Verena Fritz, Defne Gencer, Nora Kaoues, Tom Kerr, Jose Antonio Cuesta Leiva, Dena Ringold, Stephane Straub, Stephen Stretton, Richard Sutherland, William Sutton, Ioannis Vasileiou, Andrea Fitri Woodhouse, and Nkulumo Zinyengere.

Mary C. Fisk of the World Bank's Publishing Unit was the production editor. We are also grateful to Lucy Southwood for her skillful developmental editing of the manuscript and to Honora Mara for her thorough copyediting. Bill Pragluski, Critical Stages, was the designer. Communications support was provided by Carl Hanlon, Ferzina Banaji, Melissa Bryant, Catherine Sear, Jessica Brand, Joana Lopes, Alexis Condon, Sarah Farhat, and Jon Race.

The report was prepared under the guidance and leadership of Juergen Voegele, Vice President for Sustainable Development. Finally, the team acknowledges the generous support of the Climate Support Facility and guidance from Jennifer Sara, Director of the Climate Change Group, and Renaud Seligmann, Director for Strategy and Operations of the Sustainable Development Practice Group.

About the Authors

Paolo Avner is a senior economist with the World Bank and the Global Facility for Disaster Reduction and Recovery, where he co-leads the Disaster Risk Analytics and Resilient Infrastructure thematic areas. His current work focuses on the links between urban form, land uses, transportation systems, labor markets, and vulnerability to natural hazards in developing country cities. He has worked on a number of analytical products, including Urbanization Reviews and flagship reports and is the author of several policy-oriented research papers. Before joining the World Bank, he worked in France as a researcher at LEPII (Grenoble) and the Center for International Research in Environment and Development (Paris). He graduated from Sorbonne University and Université Paris X Nanterre as an economist and holds a PhD from École des Hautes Études en Sciences Sociales (Paris).

Ira Irina Dorband is an economist at the World Bank. Her research applies quantitative and qualitative methods from macroeconomic modeling to expert interviews to analyze the interplay between climate and development policies, focusing on equity and political economy aspects. Her work focuses on developing novel global models and applied tools to provide policy support on the opportunities of low-carbon development for industries, workers, and consumers. Before joining the World Bank in 2020, she worked in climate policy research and consulting in Africa, Asia, and Europe. She is a PhD candidate at the Mercator Research Institute of Global Commons (MCC Berlin) and the Technical University of Berlin, and she holds a master's degree from Duke University and the Free University of Berlin.

Catrina Godinho is a political economist working on topics related to climate and development. She is a fellow of the Energy for Growth Hub and an Agora Energiewende alumna, and was previously a senior research fellow at the Oxford Institute of Energy Studies. Her research interests include the political economy of structural and policy reforms; sustainable development; climate policy, action, and tracking; just transitions; and energy development and governance. She is an author of several articles, working papers, reports, and book chapters. She holds a master's degree in social science from the University of Cape Town, South Africa.

Stéphane Hallegatte is a senior climate change adviser at the World Bank. He joined the World Bank in 2012 after 10 years of academic research in economics and climate science. His research interests include development and poverty economics; the economics of natural disasters, risk management, and climate change adaptation; urban policy and economics; and low-carbon development and green growth. Stéphane was a lead author of the Fifth Assessment Report of the Intergovernmental Panel on Climate Change. He is an author of dozens of articles published in international journals in multiple disciplines and of several World Bank books, including *Shock Waves:*

Managing the Impacts of Climate Change on Poverty (2016), *Unbreakable: Building the Resilience of the Poor in the Face of Natural Disasters* (2017), and *Lifelines: The Resilient Infrastructure Opportunity* (2019). He has an a master's degree in meteorology and climatology from Université Paul Sabatier (Toulouse) and a PhD in economics from École des Hautes Études en Sciences Sociales (Paris).

Camilla Knudsen is an economist at the World Bank, where she works to integrate climate change and development considerations. Most recently, her work focuses on the distributional impacts of climate policies in terms of employment outcomes for workers. Before joining the World Bank in 2021, she completed a PhD in environmental economics at the University of Manchester, United Kingdom. Her research focused on nonmarket valuation and ecosystem services.

Jana Lemke joined the World Bank in 2022 under the German Carlo Schmid Program. She works as a consultant in the Climate Change Unit, looking at the links among climate change, environmental degradation, and marginalization. Before joining the World Bank, she supported the project management of climate change adaptation and mitigation, as well as ecosystem protection projects in Central Africa and Southeast Asia. She holds a master's degree in international development studies from the Philipps-University of Marburg, Germany.

Penny Mealy is a senior economist at the World Bank, a research associate at the Institute for New Economic Thinking and the Oxford Smith School of Enterprise and the Environment, an adjunct senior research fellow at SoDa Labs at the Monash Business School, and an external applied complexity fellow at the Santa Fe Institute. Her work applies various methods from complex systems and data science to analyze the interrelated challenges of climate change and economic development. Her research has developed novel, data-driven approaches for analyzing structural change, occupational mobility and the future of work, and the transition to the green economy. She has held various research fellow roles at the Oxford Martin School; the Oxford Smith School of Enterprise and the Environment; the Bennett Institute for Public Policy at Cambridge University; and SoDa Labs, Monash University. She has also frequently advised international organizations, governments, and businesses on green growth and development strategies. She completed a PhD at the Institute for New Economic Thinking at Oxford University.

Jun Rentschler is a senior economist at the World Bank, working at the intersection of climate change and sustainable resilient development. Before joining the World Bank in 2012, he served as an economic adviser at the German Foreign Ministry. He also spent two years at the European Bank for Reconstruction and Development working on private sector investment projects in resource efficiency and climate change. Before that, he worked on projects with Grameen Microfinance Bank in Bangladesh and USAID's Partners for Financial Stability Program in Poland. He is a visiting fellow at the Payne Institute for Public Policy, following previous affiliations with the Oxford Institute for Energy Studies and the Graduate Institute for Policy Studies in Tokyo. He holds a PhD in economics from University College London, specializing in development, climate, and energy.

Main Messages

By adopting the Paris Agreement in 2015, 195 governments agreed to hold global warming at well below 2°C above preindustrial levels and pursue efforts to limit it to 1.5°C. Despite multiple pledges and commitments, the rapid progress in key technologies, and the many policies introduced to date, the world is not on track to meet this objective. Moreover, despite robust evidence that countries have opportunities to reduce emissions at no or even negative costs, the failure to seize these opportunities suggests that the main obstacle is neither economic nor technological. Rather, the political economy is proving to be the key barrier to progress.

This obstacle is not impassable: there are many examples of successfully implemented climate policies. For example, in 2014, defying political and economic challenges, the Arab Republic of Egypt's Energy Subsidy Reform eased fiscal pressures and encouraged greater private investment in clean energy, with solar and wind generation growing almost threefold in the following five years. In Canada, the province of British Columbia introduced a carbon tax after the financial crisis in 2008, covering 70 percent of greenhouse gas emissions. That reform has reduced emissions and inequality, has raised growth and employment, and now has the support of a majority of citizens. Kenya reformed its power sector, a sensitive and important source of revenues and influence, thus improving efficiency, increasing cost recovery, and mobilizing private sector investments into renewable energy. The lesson from these case studies is clear—climate action with an impact is possible in the real world.

This book sets out why climate policies are successfully adopted in some cases but meet substantial opposition in others. Guided by the 4i Framework—covering four key components of the political economy: *institutions, interests, ideas,* and *influence*—it offers a framework to help policy makers replicate these achievements and effectively maneuver through a multitude of political economy barriers.

Climate change presents a unique challenge in that policy makers need to balance the speed and scale required to achieve global climate objectives with the time required to ensure political acceptability and social sustainability. To implement sustainable and transformative climate policies, policy makers can approach the design of their climate strategy and policies along four dimensions.

1. **Climate governance: strategically adapt the institutional architecture and embed climate objectives into a positive development narrative.** Institutions frame the relationship between actors and shape their influence, ideas, and interests. Policy makers can start by strategically using and adapting the institutional context for the climate transition, for instance, through climate change legislation, long-term strategies, or just transition frameworks.

   If climate change mitigation objectives are already widely recognized in public debates and polarization on climate policy is low, policy makers can build strategic climate institutions to help mediate interest groups and build consensus, facilitate and inform stakeholder engagement and alignment, foster supportive coalitions, and improve the overall institutional context. As demonstrated by the European Union's Fit-for-55 plan, these kinds of institutions can create stability and predictability, reduce the likelihood of policy reversal, and help maintain a consistent and cost-efficient strategy, even if the political context changes.

If climate change mitigation objectives are less consensual, governments can layer climate governance functions into existing institutional structures and policy objectives. For example, proactive climate-oriented entities have emerged within various Indian government ministries, achieving tangible outcomes by incorporating a climate perspective into existing organizational frameworks and aligning it with established priorities. Between 2014 and 2022, India's renewable power generation capacity more than doubled, while energy efficiency improved, enhancing energy security.

2. **Policy sequencing: balance short-term feasibility and long-term ambition.** Because the political economy and institutional contexts are not static, policy makers need to follow a dynamic approach in designing and implementing reforms. Policy prioritization can be based not only on technical and political feasibility but also on the ability to actively build political support, increase capacity, and reduce the costs of future climate action. For example, policies that create interest groups that benefit from and support climate action can facilitate and enable further action, such as in China, where industrial policies in the mid-2000s supported renewable energy industries, thereby paving the way for the successful launch of an emissions trading scheme in 2017.

   Because climate policy adoption is path-dependent, it is much easier to introduce policies that build on existing institutional capacity and know-how. With the help of the Climate Policy Feasibility Frontier—a tool to inform policy choices by considering existing and expected policy-making capacity—analysis finds that, for Türkiye, a legally binding climate strategy or binding emissions reduction target and an emissions trading scheme or a carbon tax would be feasible and most likely to build momentum toward further action.

   Targeting tipping points—that is, rapid changes in social, technological, and political domains—through shifts in societal values and behavior, technology maturity and accessibility, or support for and implementation of policies can also help governments incentivize rapid and systematic change. Thanks to these tipping points, a well-sequenced approach does not need to be slow, making it possible to combine political feasibility with ambition and speed.

3. **Policy design: focus on people and manage the distributional effects of climate policies.** Policy makers also need to minimize, manage, and, if necessary, compensate for the distributional impacts of policies on the poor and vulnerable as well as on interest groups, sectors, and regions. New analysis for this book finds highly heterogenous impacts of climate policies across households, with a larger variance within than across income groups. These impacts depend not only on consumption patterns but also on factors like access to electricity or public transit. Poor people who do not consume much fossil fuel and cannot access modern transportation may not experience heavy direct impacts from climate policies, but ill-designed policies can make it harder for them to transition away from biomass or to access better-paying jobs. Near-poor and lower-middle-class households, who consume more energy and are highly vulnerable to price changes, experience larger and more visible immediate impacts from policy reform. New analyses in Cape Town, Kinshasa, and Rio de Janeiro show that higher

transportation fuel costs have particularly large impacts on lower-middle-class households. While protecting the poorest and most vulnerable people is an imperative, political opposition is more likely to originate from impacts on well-organized or powerful interest groups, or from impacts that are concentrated on sectors or places that lack the resources or substitution options to adjust.

Possible and affordable tools for protecting poor and vulnerable populations include revenue redistribution and compensation, but it is important to consider practical challenges—such as lack of social protection infrastructure and of household data—to enable effective targeting and delivery. Active labor policies, reskilling programs, social protection systems, place-based policies, and green industrial policies form part of the toolbox policy makers can use to reduce concentrated impacts, facilitate the transition, and make policies more acceptable and sustainable over time.

4. **Policy process: use public engagement and communication to improve policies and their legitimacy.** Support for, or opposition to, a policy derives not only from people's interests but also from their perceptions of reform effects and the legitimacy of decision-making. Civic engagement can improve a policy's design, enhance legitimacy, foster compromise, and help identify unintended consequences early. Effective communication can make reforms more accessible to the public and increase support. In 2011, El Salvador's gas subsidy reform was met with opposition—particularly from lower-income households, although they were expected to benefit the most—driven by misinformation and mistrust in the government. As households started to benefit from the reform, however, their perceptions improved, and the reform eventually gained broad support. Experience from Indonesia also shows that opposition to fossil fuel subsidy reform is directly linked to local perceptions of corruption. When corruption is low, poor households are more than two-and-a-half times more likely to support than to oppose fuel subsidy reform. Without public trust, even well-designed, well-intentioned promises of compensation and redistribution can lack credibility. For a reform to be perceived as legitimate, it must involve transparent and participatory policy processes, and have outcomes that are desirable and acceptable for the public.

This book shows how appropriate governance frameworks, strong institutional capacity, well-designed policies with adequate compensation measures, and early engagement with all stakeholders are essential strategic elements to build consensus and momentum for transformative policies. By deploying these tools, policy makers can navigate the urgency of climate action and its political economy challenges to achieve their long-term climate goals and secure a livable planet.

Abbreviations

CCIA	Climate Change Institutional Assessment
CO_2	carbon dioxide
COP15	15th session of the Conference of the Parties to the 1992 United Nations Framework Convention on Climate Change (2009)
COP21	21st session of the Conference of the Parties to the 1992 United Nations Framework Convention on Climate Change (2015)
CPFF	Climate Policy Feasibility Frontier
C-PIMA	Climate-Public Investment Management Assessment
EIA	environmental impact assessment
ETS	emissions trading system
FIT	feed-in tariff
FYP	five-year plan
GB	Guobiao, Chinese for "national standard"
GDP	gross domestic product
GHG	greenhouse gas
INDC	intended nationally determined contributions
IPUMS	Integrated Public Use Microdata Series
LGCC	General Law on Climate Change (Ley General de Cambio Climático)
LTS	long-term strategy
$MtCO_2e$	million tonnes of carbon dioxide equivalent
NDC	nationally determined contribution
NO_x	nitrogen oxide
PDF	pollutant discharge free
PEFA	Public Expenditure and Financial Accountability Climate Responsive Public Financial Management Framework
PM	particulate matter
PPCDAm	Plan for the Prevention and Control of Deforestation in the Legal Amazon

PV	photovoltaic
RE	renewable energy
R&D	research and development
SO_2	sulfur dioxide
tCO_2e	tonnes of carbon dioxide equivalent
TCZ	two control zones
TWh	terawatt hours
UNFCCC	United Nations Framework Convention on Climate Change

Overview

Navigating politics: A key obstacle to urgent climate action

Good intentions, yet insufficient progress

In 2015, through the Paris Agreement, 195 governments agreed to hold global warming at well below 2°C above preindustrial levels and pursue efforts to limit it to 1.5°C. The subsequent 2021 Glasgow Pact affirms the need to reduce global carbon dioxide (CO_2) emissions to net zero by midcentury to limit warming to 1.5°C. *Net zero* means reducing emissions as close to zero as possible and compensating for the remainder with carbon removals through natural carbon sinks and technological solutions. Country climate commitments signal a willingness to act and meet these goals. As of March 2023, 172 countries had submitted a new or updated nationally determined contribution (NDC) in line with the Paris Agreement's ratcheting mechanism. In addition, over 70 countries—covering 76 percent of global emissions—have pledged to reach net zero. According to Climate Action Tracker (NewClimate Institute 2022), current NDC targets and long-term pledges would lead to warming of 2°C in 2100, close to the Paris Agreement's objective.

Nevertheless, ambitious goals and aspirational commitments have not translated into the required national actions, as illustrated by the recent synthesis report of the United Nations Framework Convention on Climate Change's global stocktake technical dialogue (UNFCCC 2023). Promises aside, if countries implement only their current climate policies, warming will exceed 2°C and could be as high as 3.4°C by 2100, with devastating impacts (NewClimate Institute 2022). Adoption and implementation of climate policies have taken place much more slowly than climate goals would dictate, creating a gap between commitments and action and raising concerns about the feasibility of policies needed to achieve climate goals.

What is holding us back?

Today, the single biggest barrier to urgent climate action is neither the lack of affordable carbon-free technologies nor the lack of resources. Continuous technological innovation is providing us with modern zero-carbon solutions in transportation, energy, and agriculture, with rapidly declining costs. Many clean technologies are already cheaper than carbon-intensive ones, even without factoring in local externalities and costs, such as air pollution and dependency on and costs of energy imports. Vast volumes spent on harmful subsidies in energy, water, and agriculture illustrate that governments have resources that they could use better for resilient, low-carbon development (Damania et al. 2023). And the

World Bank's Country Climate and Development Reports identify many opportunities for synergies between development and climate objectives (World Bank Group 2022).

Why do governments not seize these opportunities? Governments certainly face challenges related to a lack of financing or access to technologies. But one key barrier is the difficulty of designing and enacting structural change in a complex political environment that is defined by a wide range of political interest groups of varying degrees of power and influence; inconducive institutional architecture; limited government capacity; and diverging preferences, views, and beliefs across people, sectors, and groups (Peng et al. 2021). This complex context in each country—known as the *political economy*—ultimately enables or constrains effective responses to the threat of climate change.

There is no time to lose

The conventional approach in such contexts is to patiently work with and around these political economy constraints. This approach means gradually reforming institutions, carefully monitoring public support, and compromising on the speed and scale of policy implementation in favor of maximizing consensus building and minimizing disruption. But climate change is unique in that it requires urgent transformative action that can be delayed no longer.

Achieving this transformation is easier said than done. No technological innovation can magically solve the many political economy barriers to enacting climate policy, and no single approach can successfully overcome these barriers in all countries and contexts. Policy makers will have to find the right compromise between urgent action and taking the time needed to ensure political and social acceptability and sustainability.

The political economy can be managed, and change can happen fast

Countries have not been idle: they have implemented climate policies that are curbing emissions growth. High-visibility failures and unrest, as recently seen in Ecuador and France, hide a large number of climate policies that are being successfully implemented. According to the Climate Policy Database, countries have announced more than 4,500 climate policies over the last three decades.[1] This book's companion report, *Reality Check: Lessons from 25 Policies Advancing a Low-Carbon Future*, provides examples of successful implementation of climate policies, even in difficult political economy contexts (World Bank 2023). Many of the examples are neither first-best policies nor best practice. Faced with institutional capacity constraints and the need to manage trade-offs with other policy objectives, governments have often had to compromise. But these interventions have managed to draw enough support to be successfully implemented, and to create momentum toward more climate action.

- *Costa Rica's National Decarbonization Plan*, one of the most ambitious global strategies for low-carbon development for a middle-income country, helped align stakeholders' expectations and mobilize at least US$2.4 billion in international concessional finance.
- *The Arab Republic of Egypt's Energy Subsidy Reform* offers an example of successful fossil fuel subsidy reform—a notoriously challenging policy to implement. It has eased fiscal pressures and has encouraged greater private investment in clean energy, with solar and wind generation growing almost threefold between 2014 and 2019.
- *The carbon tax in the Canadian province of British Columbia*, introduced in 2008 right after the global financial crisis and covering 70 percent of greenhouse gas emissions,

has reduced emissions and inequality, raised growth and employment, and now receives majority support from citizens.

- *Climate-smart agriculture in Africa's Sahel region,* thanks to targeted interventions, has been adopted by farmers in the form of low-cost, efficient traditional practices. For example, in Niger, farmer-managed natural regeneration increased yields by 16–30 percent between 2003 and 2008 and added nearly 5 million hectares of tree cover.
- *With the help of Colombia's mandatory green building code enacted in 2015,* 11.5 million square meters of space built or under construction got certified as green under the International Finance Corporation's EDGE program by the end of 2022, and 27 percent of new buildings were certified between 2021 and 2022.
- *India's national solar mission* has made the country one of the world's most rapidly growing solar markets, with solar growing from 4 percent to 13 percent of power generation between 2014 and 2022. The private sector has been heavily involved, investing US$130 billion since 2004.
- *Kenya's power sector reforms* have made it one of the most successful countries in attracting private financing for clean power assets and Africa's largest developer of geothermal power. Since 2000, the CO_2 intensity of power generation has fallen fourfold. As well as reducing greenhouse gas emissions, these reforms have made electricity supply more reliable and increased people's access to energy.

This book aims to understand why these measures were successfully implemented, whereas similar measures have triggered strong opposition elsewhere. It provides a framework to help policy makers reproduce these successes in other contexts and successfully navigate political economy constraints (see box O.1 for what this book does not address).

BOX O.1

What this book does not cover

An important caveat is that this book has a limited scope. In particular, it does not discuss how international agreements can create incentives for policy makers and countries to reduce greenhouse gas emissions or the dynamics through which they facilitate and enable national action. The Paris Agreement does not include a mechanism to solve the free rider problem and lacks an enforcement mechanism (Keohane and Oppenheimer 2016), but it does support "catalytic cooperation," with early movers reducing costs for others and enabling them to commit to increasingly ambitious targets over time (Hale 2020). This book does not discuss how to manage historical responsibility for future climate change, the unequal carbon budget available to lower-income countries, or the need for quicker emission reductions from high-income countries. It does not cover the weaknesses of the Paris Agreement regarding the so-called means of implementation, or the financial and technological support provided by higher-income countries to help lower-income countries adapt to and mitigate climate change, including their failure to deliver on their own commitment to "mobilize US$100 billion per year for the needs of developing countries in the context of meaningful mitigation action" (UNFCCC 2023). Although these questions are related to domestic political economy challenges, because political feasibility at the country level depends on the fairness of global processes and the fair contribution of all countries, we do not focus on these international challenges here.

Instead, we focus on *political economic barriers* to domestic policies and interventions that contribute to decarbonization. Acknowledging that policy makers have diverse interests—influenced by their responsibilities, relations to other stakeholders, exposure to lobbying, and other factors (Persson and

(Continued)

BOX O.1

What this book does not cover (continued)

Tabellini 2002)—this book focuses on how governments with a *genuine commitment to their stated decarbonization goals* can achieve them and capture identified opportunities for synergies between development and climate objectives, while juggling other objectives and managing a complex political economy context.

This book does not discuss the political economy challenges faced by adaptation and resilience goals, even though they are as important and connected to the rest of the climate challenge. The World Bank is developing a broader work program on the political economy of climate action, of which this book is only one contribution. Including further analytic work on governance, institutions, and political economy aspects that matter for climate change mitigation and adaptation, this program will support an increasing range of country engagements on various governance aspects of climate change to contribute to and support effective policy actions and enable public and private investments to secure a livable planet.

A way forward: A dynamic approach to the political economy

The political economy is not set in stone: although policy makers must consider it when designing institutions and policies, they can also change it over time. This dynamic lens leads to the following four key messages.

First, governments should aim to move from opportunistic or unstable to enduring and strategic climate institutions and to embed climate objectives into a positive development narrative. Country-level political economy dynamics—specifically climate policy narratives and political polarization—determine the best approach to climate institutions in different countries. But climate institutions that are a "good fit" for the political economy today can pave the way for more strategic climate institutions tomorrow. Climate change framework laws, long-term strategies, and just transition frameworks are three key strategic climate institutions that can fundamentally alter the political economy of climate policies, and they have been implemented in countries as diverse as Costa Rica, South Africa, and the United Kingdom.

Second, governments can prioritize policies that are feasible in a given political economy context, but that also transform the political economy by building greater political support and reducing the costs of climate action over time. This book offers tools for countries to strategically select and sequence policies that build institutional capacity, create winners who will support further policy action, or offer firms and people affordable options to substitute for fossil fuels. Governments can also leverage reinforcing policy feedback processes and target tipping points in the broader socio-technical-political system. They triggered such a tipping point by making solar power the cheapest option to generate electricity, and they can now do the same with electric vehicles or heat pumps by making them the default (and affordable) option for consumers. Thanks to such tipping points, strategically selecting and sequencing feasible policies does not mean climate progress will be slow.

Third, the design and implementation of policies need to consider the political economy, including concentrated distributional impacts and the need for policy legitimacy. Climate policies have heterogenous distribution implications across societal groups, income classes, sectors, occupations, or space, and impacts vary more within than across income groups. It is possible and affordable to protect poor and vulnerable populations through compensation; however, the political economy involves more than distributional impacts, and protecting poor households is not enough to ensure acceptability. Opposition to a policy reform is often triggered by concentrated impacts on well-organized or well-connected groups—such as powerful interest groups, organized workers in key sectors, the urban lower-middle class, carbon-intensive regions, or other societal groups—making complementary policies and compensation more challenging to design and implement.

Fourth, opposition also often originates from a perceived lack of legitimacy of (or agency in) the policy process. Civic engagement and communication can help design better policies and identify unintended consequences as soon as possible. They also help build legitimacy and develop working compromises and necessary support by mediating distributional conflict, differences in preferences and priorities, and unequal power dynamics.

The defining features of every political economy setting

At first glance, the political economy barriers that impede climate change policies may appear so wide and varied that they defy any systematic definition, let alone strategy. Governments manage many competing demands with constrained resources, and future climate change impacts are less visible and salient than immediate transition costs. Short political mandates can undermine action on long-term objectives; misinformation and lobbying can distort public discourse and opinions; high-level climate ambitions are not always in tune with the day-to-day priorities of political leaders and communities; and repeated crises can change short-term priorities overnight. There can also be considerable institutional and political barriers, sometimes linked to low institutional capacity and poor governance, and sometimes erected by vested interest groups. And political, cultural, or ideological beliefs can reinforce opposition to action.

The 4 i's: Institutions, interests, ideas, and influence

Nevertheless, governments have a long history of operating within this reality of political economy barriers. Based on decades of experience ranging from education policy to macrofiscal management, extensive evidence documents the drivers of political economy barriers and solutions. Drawing on a vast range of political economy theories, concepts, case studies, definitions, and analysis tools, this book highlights how it is possible to analyze and understand seemingly insurmountable political complexity and turn it into a guiding framework for devising effective policy strategies. Although distributional impacts have often attracted the most attention, a systematic approach can help dissect political economy issues into four key components, which make up the 4i Framework (Godinho, Hallegatte, and Rentschler, forthcoming, and figure O.1):

1. *Institutions*: the formal and informal rules, norms, and organizations that provide incentives and constraints for economic, political, and social behavior in society
2. *Interests*: heterogenous distributional impacts, as well as differences in priorities and preferences, that shape all actors' behavior
3. *Ideas*: the beliefs, values, and worldviews that shape actors' preferences
4. *Influence*: the power, authority, and leverage that actors can use to advance their interests and ideas, and their interactions with each other and institutions.

These components are the result of a long process of historical, political, social, cultural, and economic development, making them highly specific to place and time. Some factors are determined by centuries of sociopolitical history, such as the formation of

FIGURE O.1. **The 4i Framework and an iterative approach to climate policy**

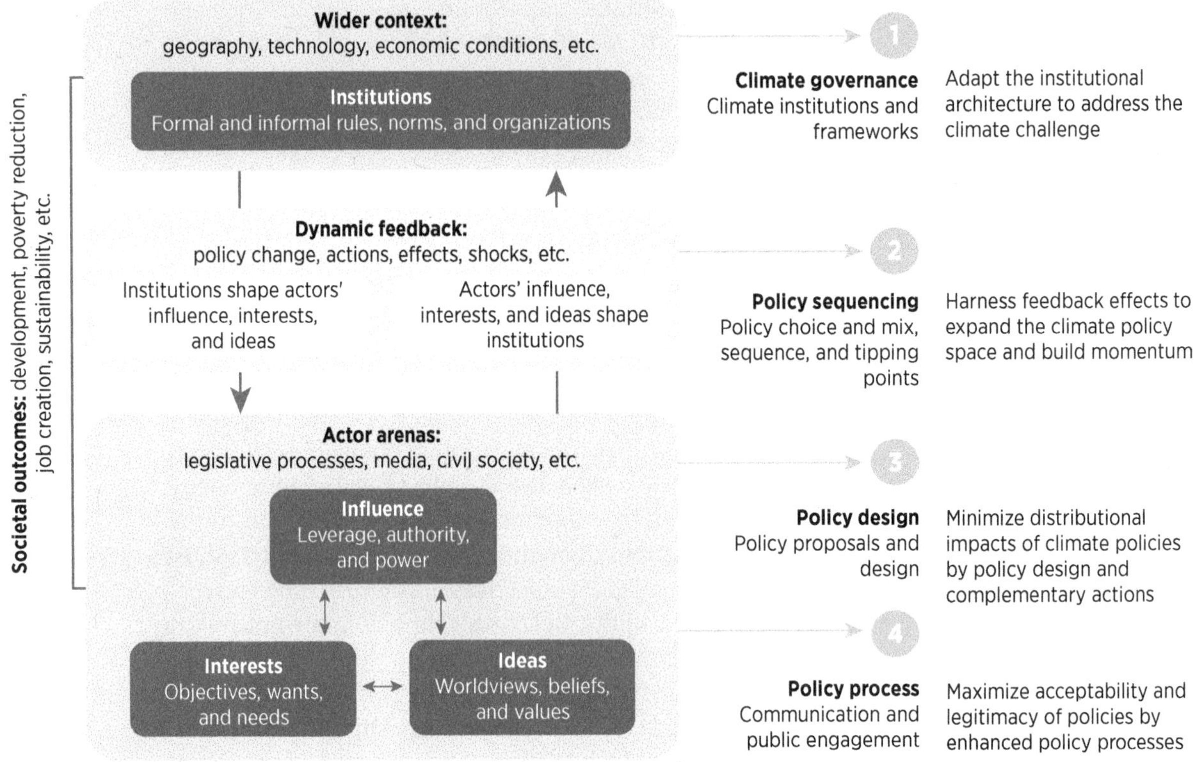

Source: Based on Godinho, Hallegatte, and Rentschler, forthcoming.

laws, values, and norms. Others are more transient, determined by an ongoing crisis or lobbying by certain interest groups. Yet they all act together to give rise to countries' institutional structures, which ultimately constitute the policy-making environment for all political and economic development, including climate change mitigation. This also means that institutions, interests, ideas, and influence can act as enablers for, or barriers to, good climate policies.

Successfully navigating the political economy for climate policies means recognizing and leveraging the enabling factors—while also avoiding or purposefully changing the barriers. Despite the attention paid to distributional impacts, they are only one dimension of the political economy, especially when winners of the policies feel like losers (Calvo-Gonzalez, Cunha, and Trezzi 2015). As illustrated by many examples discussed in this book, such as the reform of gas subsidies in El Salvador in 2011, perceptions and ideas also matter.

A four-pronged iterative approach for urgent climate action in a complex political economy

In practice, policy makers can approach the design of their climate strategy and policies along four dimensions (see figure O.1), which correspond to chapters of this book:

1. *Climate governance: strategically adapt the institutional architecture and embed climate objectives into a positive development narrative (chapter 2).* Because institutions frame

the relationship between actors—and their influence, ideas, and interest—policy makers can start by strategically using and adapting the institutional context for the climate transition, building on existing structures and societal goals.

2. *Policy sequencing: balance feasibility and long-term ambition (chapter 3).* The political economy and institutional context are dynamic and can be influenced by policies. Hence, policy makers can select their priorities, not only to make policy implementation feasible but also to actively build capacity and change the political economy and institutional context, building momentum toward the long-term objective and transformation.
3. *Policy design: focus on people and manage the distributional effects of climate policies (chapter 4).* After selecting policy priorities, policy makers need to minimize and manage the distributional impacts of policies, including impacts not only on the poor and vulnerable but also on well-organized interest groups, sectors, or regions.
4. *Policy process: use public engagement and communication to improve policies and their legitimacy (chapter 5).* Civic engagement can improve a policy's design, enhance legitimacy, foster compromises, and help identify unintended consequences early. Effective communication can make reforms more accessible to the public, increasing support and sustainability.

Climate governance: Strategically adapt the institutional architecture and embed climate objectives into a positive development narrative

To tackle complex development challenges, including climate change, governments need to organize their institutional architecture in line with national priorities, through legal and regulatory frameworks, institutions with enforcement capacity, and ministries with well-defined roles and responsibilities. In the context of mitigating climate change, the way countries organize themselves—known as *climate governance*—defines their available policy-making options. Formal climate governance institutions provide the rules, norms, and procedures that guide target setting, policy making, and implementation.

To change the political economy, climate governance institutions must first fit in

Real world examples of efforts to build climate institutions demonstrate how the political economy can shape the form and functionality of these institutions (Dubash et al. 2021). Climate-related political narratives fall between two extremes: *mitigation-centric narratives* emerge where climate change mitigation is already a well-established, high-priority public goal that allows for explicit emissions reduction framing; *embedded climate narratives* are likely when climate is lower on the agenda, and it is easier to subsume climate goals under other objectives, such as green growth, energy security, or job creation (table O.1). At the same time, countries have different levels of polarization related to climate change, and that polarization influences the stability of institutions and policies.

In places with a consensus on climate action and low levels of polarization on climate policy, governments can build *strategic climate institutions* to help mediate interest groups and build consensus, facilitate and inform stakeholder engagement and alignment, foster supportive coalitions, and improve the overall institutional context. For example, the European Union's Fit-for-55 initiative supports the alignment of member countries' policies and legislations with the EU climate objective of reducing its emissions by at least 55 percent by 2030, thereby enabling the provision of a coherent and

TABLE O.1. **Types of climate governance**

Interests		Ideas	
		Dominant narrative on climate policies	
		Embedded	Mitigation-centric
Extent of political polarization of climate policy	Low	Under-the-radar climate politics Opportunistic climate institutions	Climate consensus politics Strategic climate institutions
	High	"Contested sector" politics Unstable sectoral institutions	In-the-crossfire politics Unstable climate institutions

Source: Based on Dubash 2021.

balanced framework for climate action. Such institutions can create stability and predictability, make policy reversal less likely (though not impossible), and help maintain a consistent and cost-efficient strategy over time, even if the political context changes.

Climate institutions prematurely established in contexts without climate consensus and where climate politics are contested can trigger negative feedback effects, increasing polarization and opposition to the institution itself and to climate action more broadly. In this context, governments can instead start with *embedded institutions* by layering climate governance functions into existing institutional structures, embedding climate into other political priorities with a focus on "win-wins" or synergies. In India, active but opportunistic climate institutions have emerged across multiple ministries through the layering of a climate lens into existing bodies. Building on existing priorities—including increasing energy supply and security, and controlling air pollution—these institutions have developed, crucially, without strong national mitigation-centric strategic institutions, such as a climate change framework law, which could have triggered backlash (Pillai and Dubash 2021). And they have delivered results: Between 2014 and 2022, India's renewable power generation capacity, including hydropower, more than doubled, with solar power growing from 3 to 63 gigawatts, and energy efficiency improvements preventing 12 percent of additional annual energy use by 2018.

Climate governance frameworks that reflect societal goals can bridge today's political economy with the future

Climate laws, green growth strategies, just transition, and other climate governance frameworks can help bridge today's political economy with a net zero future by reflecting societal goals, priorities, and imperatives and linking them to climate action and outcomes. Countries have implemented the following three key tools:

1. *Climate change framework legislation* can help countries design their own effective and comprehensive strategy, setting targets and creating institutions to engage and coordinate stakeholders. Established legal frameworks also help citizens hold their governments to account for their actions and inactions and serve as a binding regulatory structure at times of change in political administrations and priorities. In 2012, Mexico became the first large oil-producing emerging economy to adopt climate legislation when its parliament passed the General Law on Climate Change, or *Ley General de Cambio Climatico*. This law established an aspirational goal of reducing emissions by 50 percent by 2050.

2. *Long-term strategies* (LTSs) are a complementary approach to developing a climate governance framework. As of October 2023, 68 countries had an LTS under the United Nations Framework Convention on Climate Change, providing a realistic pathway toward long-term objectives and identifying useful milestones for shorter-term strategies and plans. LTSs have multiple roles and functions in climate governance, starting with informing political debates and choices. Costa Rica's LTS, the *Plan Nacional de Descarbonización*, uses a whole-of-economy pathway with targets and timelines for all emitting sectors to coordinate action and identify barriers (World Bank 2023). When an LTS receives widespread support or is embedded in law, it can be a powerful instrument for maintaining momentum, coordinating action across sectors, and offering a benchmark to measure progress over time.
3. Governments can either develop *just transition frameworks*, like in South Africa, or integrate *just transition principles* into other institutions, such as climate strategies and framework laws. Just transitions will vary, depending on local context and as defined through local processes, they have four common guiding principles: distributional justice (the fair distribution of costs and benefits); procedural justice (transparent and inclusive processes); recognition justice (recognition, respect, value, and right to self-determination for all actors); and structural justice (addressing and redressing institutional structures that produce or perpetuate injustice).

Policy sequencing: Balance feasibility and long-term ambition

Policy makers face hard choices between focusing on low-hanging fruit or investing in more challenging—but more transformational—strategies and policies. Although focusing only on easy policies may ensure action, it is unlikely to trigger the systemic changes needed to reduce emissions to zero, at least in a time frame consistent with global objectives. By contrast, focusing solely on transformational policies can lead to inaction if political forces or lack of capacity makes enactment, implementation, or enforcement of climate policies impossible. Ideally, governments should balance policies' short-term political feasibility with their contribution to long-term objectives, including through a transformation of the political economy context. They can do so by selecting policies that are feasible but that also build greater political support and reduce the costs of climate action over time, leveraging reinforcing policy feedback processes, creating interest groups that support climate action, and targeting tipping points to accelerate transformational change toward net zero.

Because policy and political processes are not static, policy packages need to evolve over time in a dynamically efficient way. As governments introduce new climate policies, they create effects that alter the broader political economy. From a political economy perspective, the lowest-cost option at one point may lead to political backlash and create higher costs in the future, whereas a more expensive policy today might shift the political economy to make more efficient policies easier to implement later. For instance, taxing polluting gasoline cars may be an efficient policy, but the measure is likely to backfire if people lack access to alternatives. Investing in public transit, subsidizing charging stations, and incentivizing the production of affordable and efficient electric vehicles will require higher investments per ton of avoided carbon emissions over the short term, but are more likely to trigger virtuous mechanics of change and help achieve long-term climate objectives.

Countries can prioritize policies that are easier to introduce and build future policy-making capacity

Climate policy adoption is path-dependent and therefore partly predictable. A new analysis of past policies using the Climate Policy Database shows that policies are much easier to introduce if they build on prior related institutional capacity and know-how (Mealy et al., forthcoming). A country would have difficulty effectively implementing vehicle or industrial air pollution standards without first having the capabilities to monitor and audit vehicle or industrial performance. Governments need to consider such path dependency in policy making when thinking about policy package design, because choices today will influence policy options tomorrow.

The Climate Policy Space provides a visual representation to better understand how policies and measures can build on each other (figure O.2). The Climate Policy Space is a network in which nodes represent climate policy instruments linked according to how often they are present together in a country. In panel a of figure O.2, policies are colored and sized according to their prevalence across countries (darker policies have been more commonly introduced); in panel b, they are colored by key policy cluster, from highly prevalent nonbinding targets and climate strategies that are fairly easy to introduce to policy instruments that relate mostly to technological deployment and innovation.

Countries with different income levels are concentrated in different regions of the Climate Policy Space. Panel c shows that low-income countries, which typically have less developed levels of institutional capacity, tend to have introduced mostly nonbinding targets or strategies. Lower-middle- and upper-middle-income countries show a broader range of policies, suggesting that expanding policy-making capacity into binding targets, institutional creation, and regulatory and market-based instruments may go hand in hand with rising levels of economic development. High-income countries span a vast range in the Climate Policy Space network, with a notable presence in technology-centric policies, arguably the actions that require the most capacity.

Combined with the usual analysis of the efficacy, costs, and benefits of policies, the Climate Policy Feasibility Frontier (CPFF) can help inform policy choices that realistically work with countries' policy-making capacity and gradually build greater capacity to introduce more ambitious types of policy. The CPFF has two key dimensions:

1. *Relative likelihood of introducing a policy in the next five years.* Based on how related a new policy is to a country's existing set of policies, this metric is expressed in relative terms, comparing policies without measuring their absolute likelihood. It measures the ease of implementing a given policy, based on a country's prior policy experience and inferred policy-making capacities.
2. *Capacity-building potential.* Reflecting the extent to which a new policy will increase the likelihood of implementing further policies in the future, this metric aims to capture the learning and capacity development potential associated with introducing a new policy and measures how introducing a given policy is expected to change a country's policy-making capacity, making it easier to implement other climate policies.

The CPFF identifies policies that may be easier to implement, as well as step-by-step pathways toward a desired policy. Figure O.3 maps Türkiye's and Viet Nam's positions in the Climate Policy Space (panels a and c) and their CPFFs (panels b and d). Each dot in the CPFFs denotes a new policy the countries have not introduced before and

FIGURE O.2. The climate policy space

a. The Climate Policy Space
Climate policy prevalence
High
Low
Removal of fossil fuel subsidies
User charges
Procurement rules
Public voluntary schemes
Unilateral commitments (private sector)
Technology deployment and diffusion
Demonstration project
Technology development
Advice or aid in inplementation
Funds to subnational governments
Information provision
Green certificates
Retirement premium
Vehicle air pollution standards
Sectoral standards
R&D funding
GHG emission reduction crediting and offset mechanism
Industrial air pollution standards
GHG emission allowances
CO2 taxes
Monitoring
Grants and subsidies
Net metering
Auditing
Infrastructure investments
Loans
Coordinating body for climate strategy
Building codes and standards
Energy and other taxes
Feed-in tariffs
Tax relief
Grid access and priority for renewables
Obligation schemes
Institutional creation
Formal and legally binding GHG reduction target
Strategic planning
Formal and legally binding renewable energy target
Political and nonbinding renewable energy target
Political and nonbinding GHG reduction target
Political and nonbinding climate strategy
Formal and legally binding climate strategy
Formal and legally binding energy efficiency target
b. Key policy clusters
Mostly technology, innovation, and miscellaneous rare policies
Mostly regulatory, market-based instruments and enabling policies
Nonbinding targets and strategy
Binding targets and institutional creation
Nonbinding GHG reduction target
c. Climate policy prevalence for countries at different income levels
Climate policy prevalence
High
Low
Low-income countries
Lower-middle-income countries
Upper-middle-income countries
High-income countries

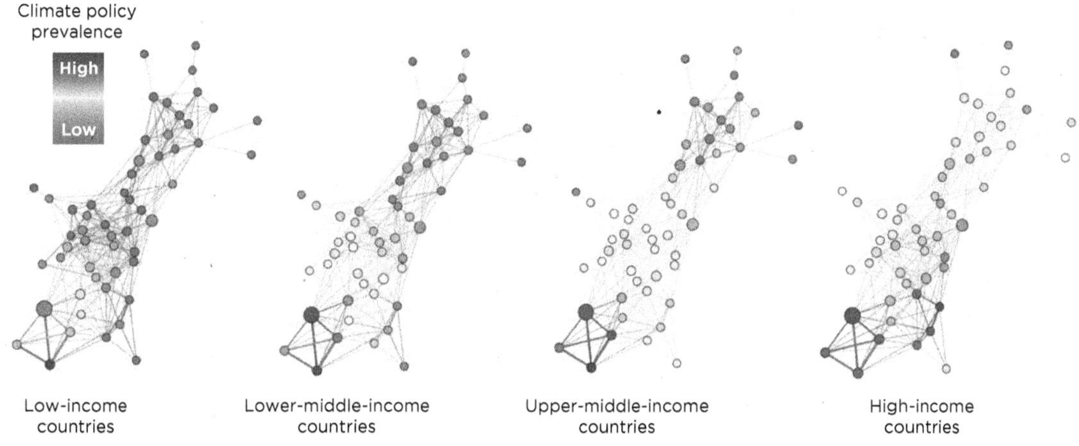

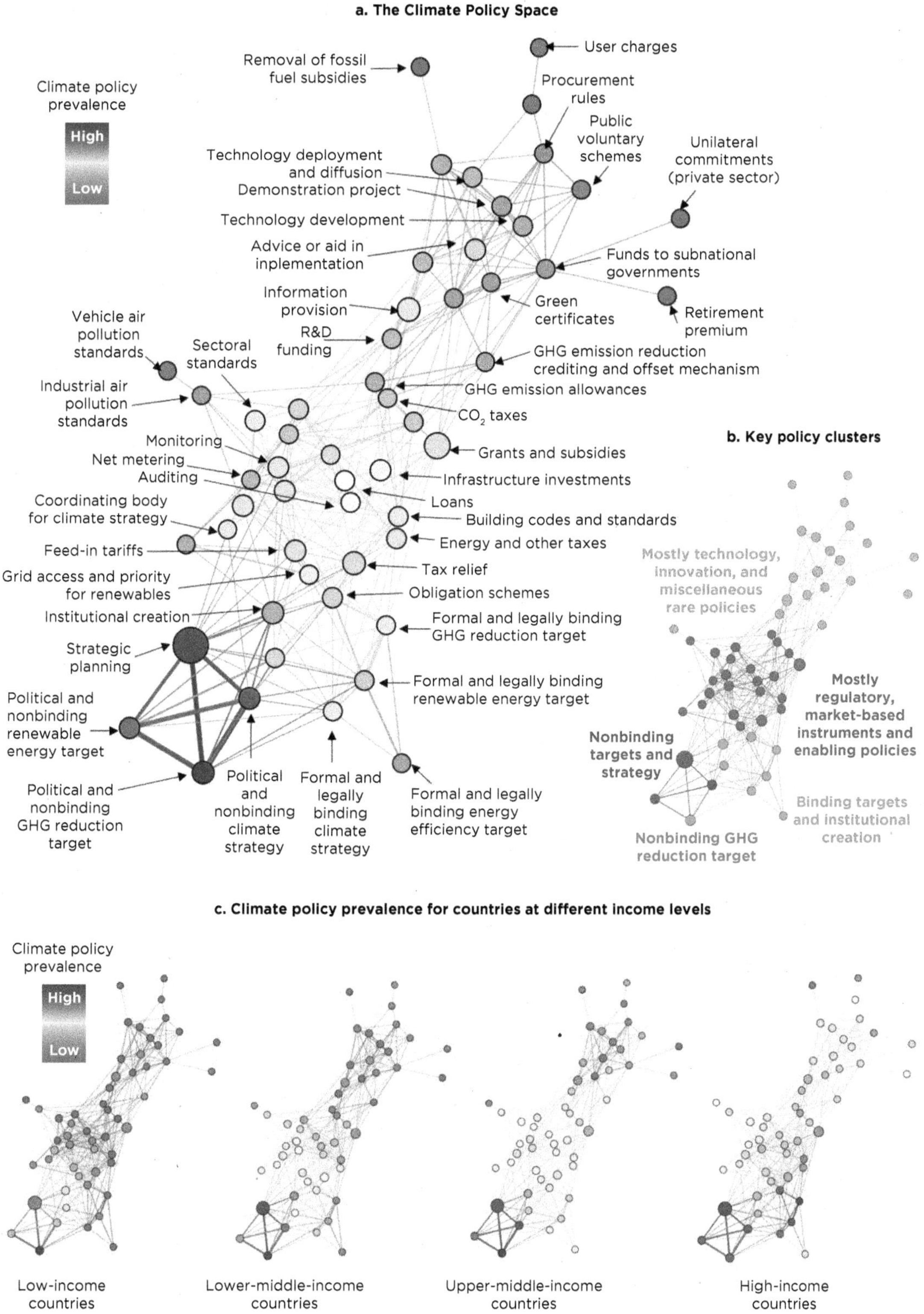

Source: Mealy et al., forthcoming.
Note: CO_2 = carbon dioxide; GHG = greenhouse gas; R&D = research and development.

FIGURE O.3. Current and feasible climate policies, Türkiye and Viet Nam

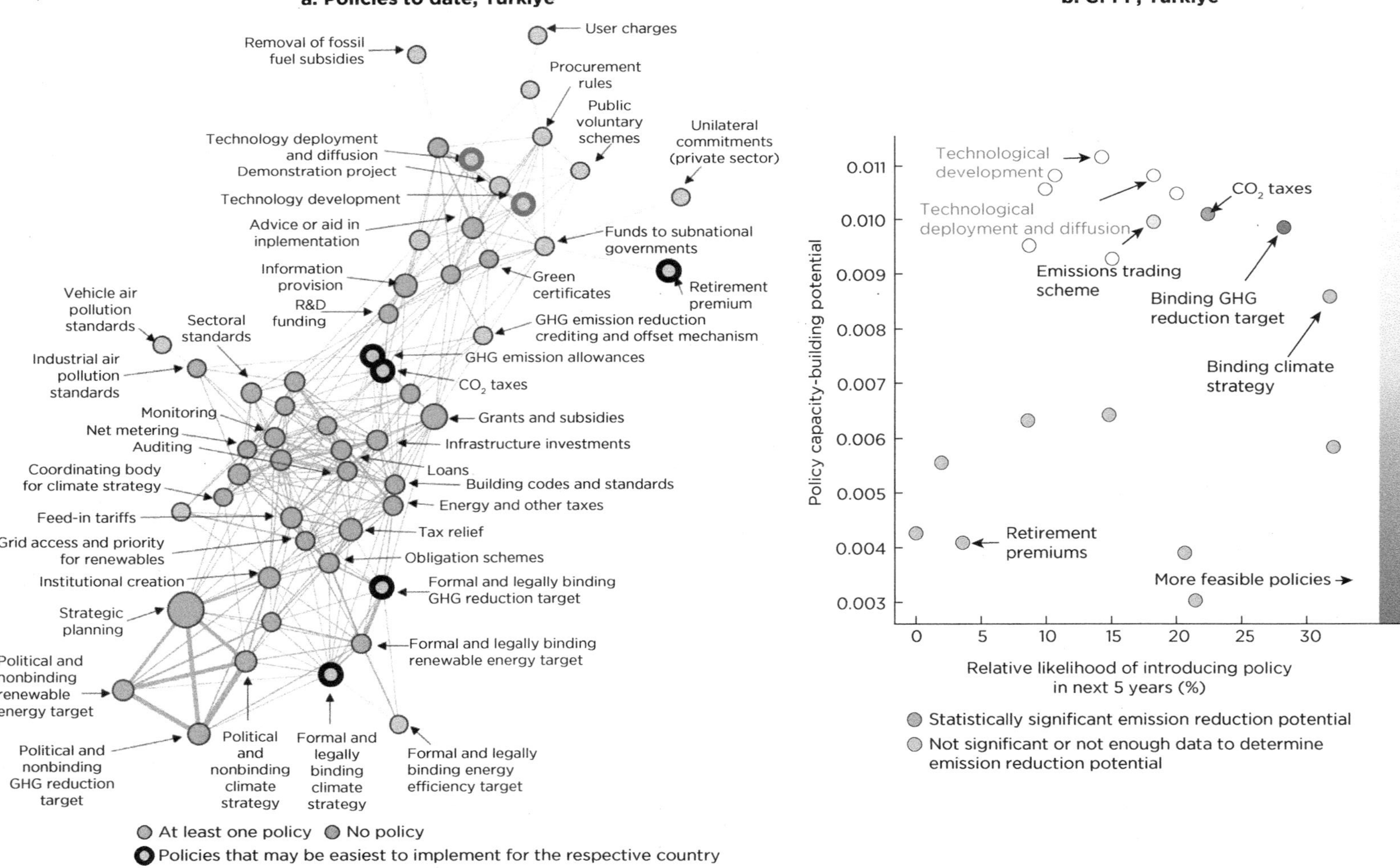

(Continued)

FIGURE O.3. **Current and feasible climate policies, Türkiye and Viet Nam** (continued)

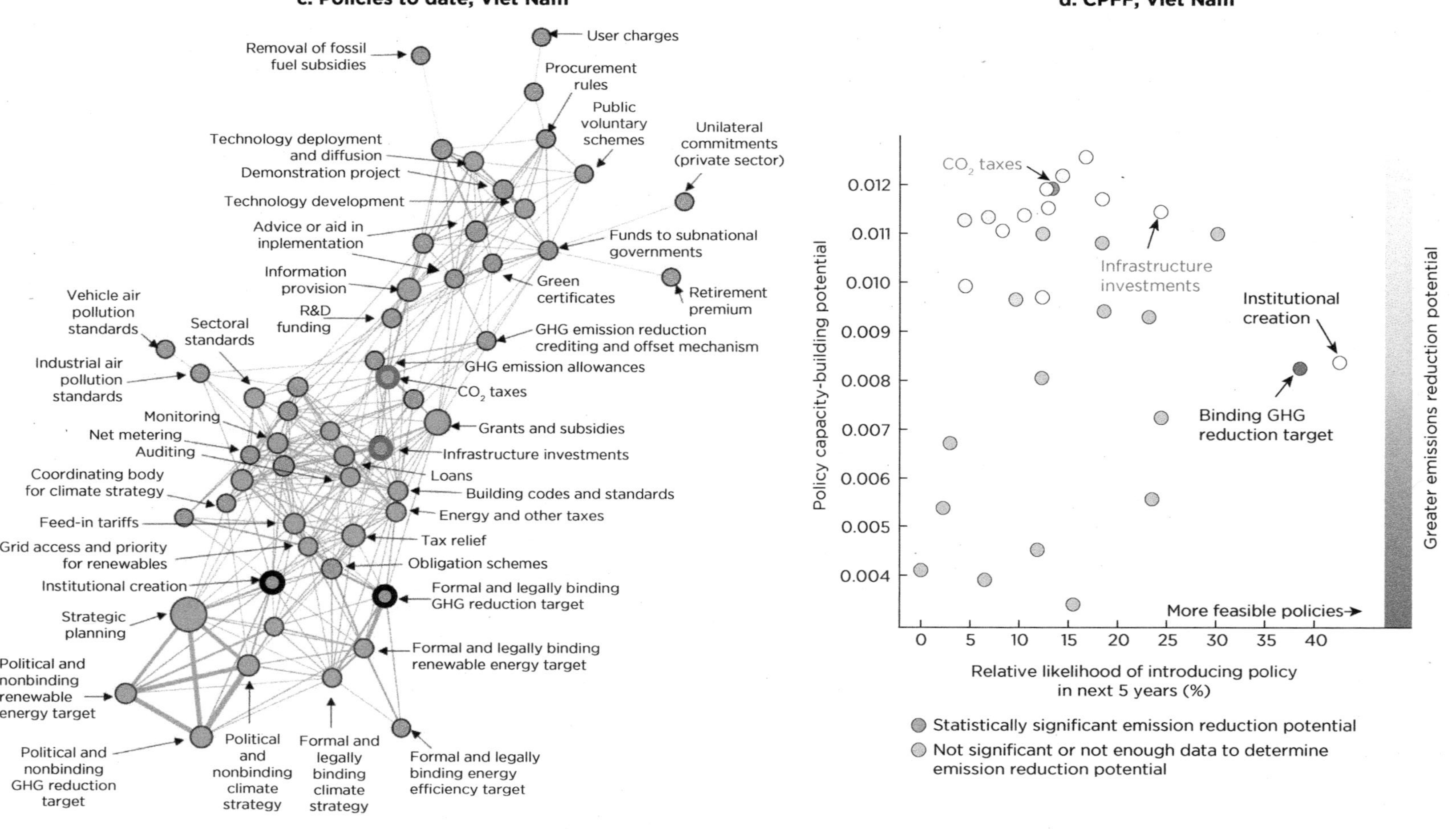

Source: Mealy et al., forthcoming.
Note: CO_2 = carbon dioxide; CPFF = Climate Policy Feasibility Frontier; GHG = greenhouse gas; R&D = research and development.

corresponds to policies colored in gray in the climate policy space. For Türkiye, policies that appear the easiest to introduce in the next five years include a legally binding climate strategy and a binding emissions reduction target, which would have to be included in a law, such as a climate change framework law.[2] Its CPFF also identifies an emissions trading scheme and a carbon tax as potential next steps. The government has recently announced its intention to develop the former, confirming it as a highly feasible intervention for the country.

The CPFF emphasizes the importance of country context and identifies different recommendations across countries. For Viet Nam, which has a different set of institutional capabilities from that in Türkiye, the most feasible and likely policies include institutional creation and binding greenhouse gas reduction targets. Policies to boost climate-related infrastructure investments could help Viet Nam build further capacity; although CO_2 taxes are likely to be less feasible in the short term, other policies can help pave the way toward this goal.

Governments can leverage reinforcing policy feedback processes and target tipping points

Strategically selecting and sequencing feasible policies to build greater institutional capacity and political support does not mean climate progress will be slow. By taking advantage of the dynamism of socio-technical-political systems, governments can build momentum to accelerate transformational climate action. Introducing specific policies can transform the associated politics, which in turn shapes the future space of policy possibilities. *Policy feedback* relates to the effects that policies can have in either reinforcing or undermining the direction or pace of future policy making. The adage "new policy creates new politics" captures the way each climate policy or intervention affects the political economy landscape, creating new incentives, spreading new ideas, supporting new coalitions, and reforming institutions.

Some policies can drive positive, reinforcing feedback effects, leading to faster climate progress and more ambitious action. For example, China's national sustainable energy policy was sequenced to reduce resistance from existing institutions and increase support by fostering winning coalitions, while gradually increasing policy stringency and reducing costs (Li and Taeihagh 2020). First, the government provided support for renewable energy, building interest groups in these technologies, before gradually ratcheting up policy stringency. Once renewable energy became cost-competitive, the government started reducing feed-in tariffs and other subsidies.

Governments can also aim to strategically target tipping points in social, technological, and political domains, which can drive rapid and systemic change. A *tipping point* refers to nonlinear change in a complex system, in terms of the speed or nature of change. The primary driver of a tipping point is the dominance of positive over negative feedback effects, which reinforce change. Several types of tipping point are relevant for accelerating progress toward net zero:

- *Social tipping points*: Rapid self-reinforcing shifts in attitudes, beliefs, behavior, and values in society
- *Technological tipping points*: Significant shifts in technology maturity, performance, costs, or accessibility
- *Policy tipping points*: Rapid shifts in support for and implementation of a particular policy or set of policies.

Sharpe and Lenton (2021) explore relative technology cost tipping points, when low-carbon technology (renewable energy in the power sector; electric vehicles in the transportation sector) becomes cheaper than high-emitting technology (coal and gas in the power sector; internal combustion engines in the transportation sector), first with—and then without—policy support. They show how these tipping points lead to changes not only in technology shares but also in climate politics, enabling more ambitious and faster policy change. Triggering one tipping point can increase the likelihood of triggering another, leading to a virtuous cycle of increasingly rapid progress on decarbonization.

Policy design: Focus on people and manage the distributional effects of climate policies

Who benefits, who loses, and who pays are not the only drivers of political economy challenges, but they remain key. Distributional conflict has two aspects. People might mobilize against a policy if they think it may negatively affect them (this includes powerful incumbents or groups) or if they think the policy has unfair or unjust impacts, for them or for society at large. To reduce the risk of distributional conflict, policy makers can do the following:

- *Reduce distributional effects* with policies that remain as neutral as possible and avoid disproportional impacts on some categories. If a climate policy creates incentives to reduce emissions, however, it is hard to avoid concentrated impacts on carbon-intensive sectors or activities, such as coal mining.
- *Compensate people or groups that are negatively affected,* either directly or indirectly. For example, Indonesia increased its social protection transfers to poor households during its fossil fuel subsidy reform. Countries sometimes target transfers toward powerful groups that are negatively affected, even without an ethical justification to do so, because these groups have de facto veto power.
- *Improve policy legitimacy* by adopting inclusive, transparent processes. People will more easily accept policies with significant impacts when they believe the process that led to the design and implementation of those policies is legitimate and just (Barron et al. 2023).

Climate policies affect different groups through multiple, intersecting channels; and distributional impacts are complex and heterogenous, crossing income groups, sectors, occupations, locations, genders, ethnicities, or other characteristics. Analyzing consumption by income classes does not capture all distributional issues, so multidimensional analyses are important for designing fair policies. These analyses are also important for managing the political economy, because organized opposition to a policy is more likely by economic sector, geographical area, or ethnic group than by income class.

Consumption effects

Because of different consumption patterns, the impacts of carbon pricing on consumption vary across income levels, but differences *within* income groups are larger than those *across* income groups (Dorband et al. 2019; Dorband et al. 2022; Douenne 2020; Feindt et al. 2021; Missbach, Steckel, and Vogt-Schilb 2022). In countries where poorer households have limited access to energy-consuming assets and services—such as cars, air conditioning, or gas for heating and cooking—these households are less exposed than richer households to an increase in fossil fuel prices. Thus, carbon pricing tends to have a lower and neutral impact in low-income countries and a larger and progressive effect in lower-middle-income countries where poor households have lower-than-average energy expenditure (figure O.4). Among upper-middle-income countries, the evidence is more mixed and varies with levels of access to public transportation and electricity and other lower-carbon alternatives. However, the near-poor and lower-middle classes, who have enough resources to consume fossil fuels but are vulnerable to small price changes, appear particularly vulnerable.

These findings may, however, underestimate the vulnerability of poor people. Because poor people tend to spend a large fraction of their income on food, they can be heavily affected by climate policies that translate into higher food prices. And, although carbon pricing systems rarely cover emissions from nonenergy sources, ill-designed climate policies that negatively affect agriculture and food systems—for example, by reducing access to key inputs—could have large impacts on food prices, and therefore on poverty. It is also important to take a dynamic view: even where higher energy prices do not affect poor people now, such policies could slow down progress toward universal access to modern energy, clean cooking, and food security. For example, an increase in fossil fuel prices may not directly affect households that cook with biomass, but the change in price may delay the ability of these households to shift to modern cooking techniques if they do not have access to affordable electricity and electric cookstoves (Greve and Lay 2023).

FIGURE O.4. **Illustration of the consumption impacts of a (noncompensated) increase in fuel prices, in a subset of countries, by income level**

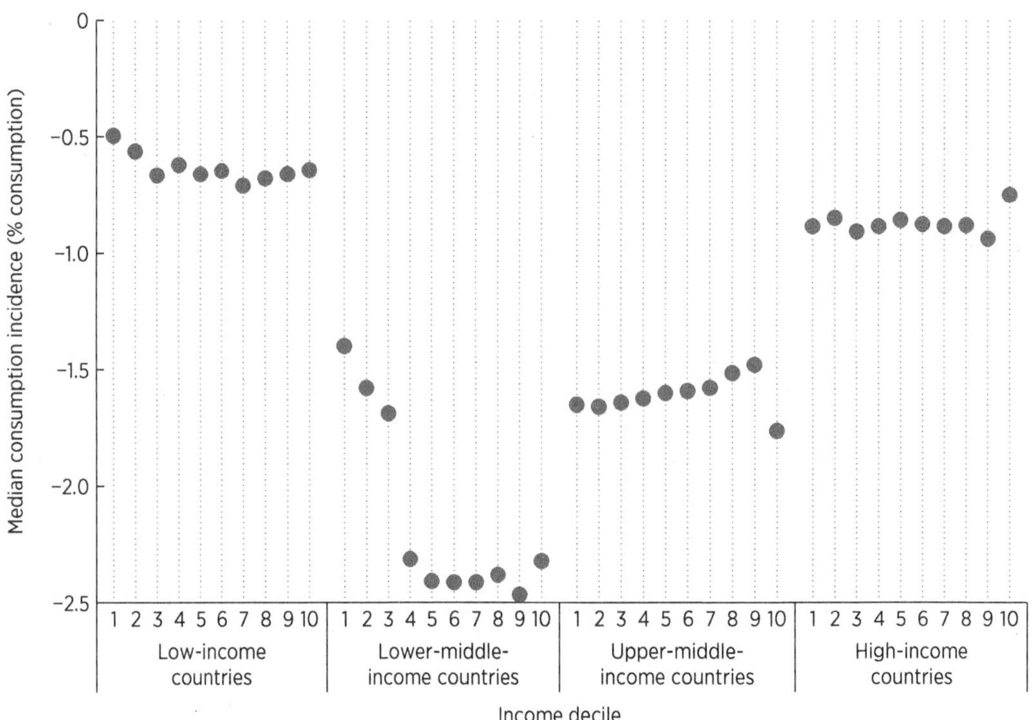

Source: Dorband, forthcoming, using the Climate Policy Assessment Tool developed by the International Monetary Fund and the World Bank to estimate the impact of carbon pricing (https://www.worldbank.org/en/topic/climatechange/brief/climate-policy-assessment-tool).

Note: These illustrative simulations, performed in 74 countries for this book, assume the introduction of a tax of US$60 per ton of carbon dioxide and the removal of energy and fossil fuel subsidies, with no recycling of the revenues or savings. These assumptions are not meant to be realistic policy packages but to illustrate the vulnerability of households to changes in fuel prices. The figure shows median impact per decile and then per country; because some households are very heavily affected, average impacts are larger than median impacts.

Employment effects

Ample evidence demonstrates that well-designed climate policies can be net job creators (Godinho 2022; Metcalf and Stock 2020; World Bank Group 2022). But such positive employment outcomes are not automatic; rather, they depend on the design of climate and other socioeconomic policies. Job creation can be driven by policy design, say, when governments recycle revenues from a carbon pricing scheme into the economy through infrastructure investments or tax breaks. It is also linked to the transition itself—for example, when climate-smart agriculture practices or renewable energy solutions are more job-intensive than existing practices and technologies. And job creation in sunrise sectors,[3] such as renewable power or electric vehicle global value chains, will depend on the investment climate, available infrastructure, a trained labor force, and appropriate tax policies and trade regulations.

To understand the effect of climate and development policies on labor and skills, a new analysis for this book explores the implications of an illustrative climate policy package combining carbon pricing with cash transfers and investments in infrastructure and public goods (Dorband, forthcoming). Despite mostly positive net effects, sectoral reallocations and policy-induced structural changes can be sizeable, particularly in more carbon-intensive economies. Jobs tend to be reallocated rather than lost, but overall job losses may be as large as 1–2 percent and gains amount to 3–4 percent of total baseline employment. And, although carbon-intensive economies may not experience larger net effects, they do undergo larger structural change, with more reshuffling of jobs.

Spatial effects

Distributional effects can emerge, even within urban areas. Climate policies that increase the price of certain transportation options can induce changes in transportation behavior and decisions about where people live and work. New evidence for this book shows a large heterogeneity in the final impacts, with a big role for adjustments through housing and labor markets. For example, fuel price increases do not have regressive impacts in Kinshasa, Democratic Republic of Congo (map O.1), because the poorest people are already priced out of—and excluded from—energy-intensive services or have limited access to areas with a high concentration of jobs (Nell et al. 2023). As a result, increased fuel prices affect poorer households less than slightly richer households, which are more likely to lose access to services or areas. Combined with the lower voice and influence of the poorest households, this higher vulnerability of near-poor households may explain why protecting the poorest alone has failed to ensure acceptability of climate or energy policies, as happened in Ecuador.

Other dimensions of exclusion and injustice

The most vulnerable members of society tend to be those who experience social exclusion and structural injustice based on ethnicity, gender, age, religion, and other factors. Ex ante assessments of consumption, employment, income, and spatial distributional effects are less able to predict these outcomes, which largely depend on political economy factors, such as institutional discrimination, culture, ideology, and unequal power dynamics (Peng et al. 2021). Ex post assessments of the social effects of climate policies therefore represent an important area of analysis that can help policy makers better manage the intersection of distributional effects and social inequality (see box O.2 for gender implications of climate policies in the agriculture sector).

MAP O.1. Losses in accessibility of jobs in Kinshasa, Democratic Republic of Congo, in a scenario with a 100 percent fuel price increase, by area and income decile

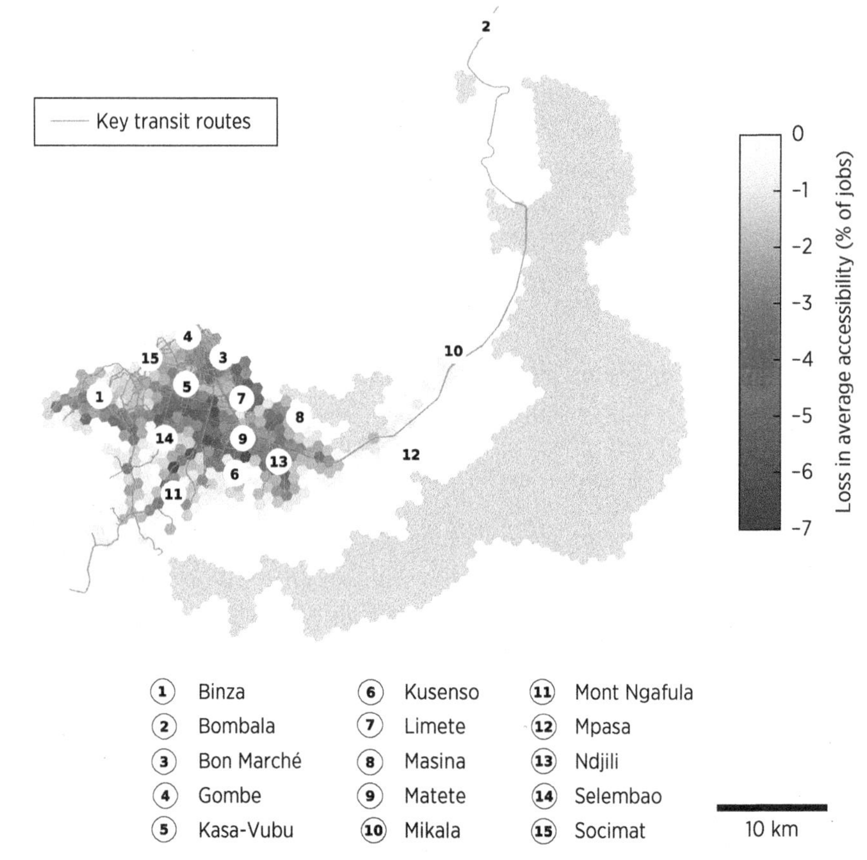

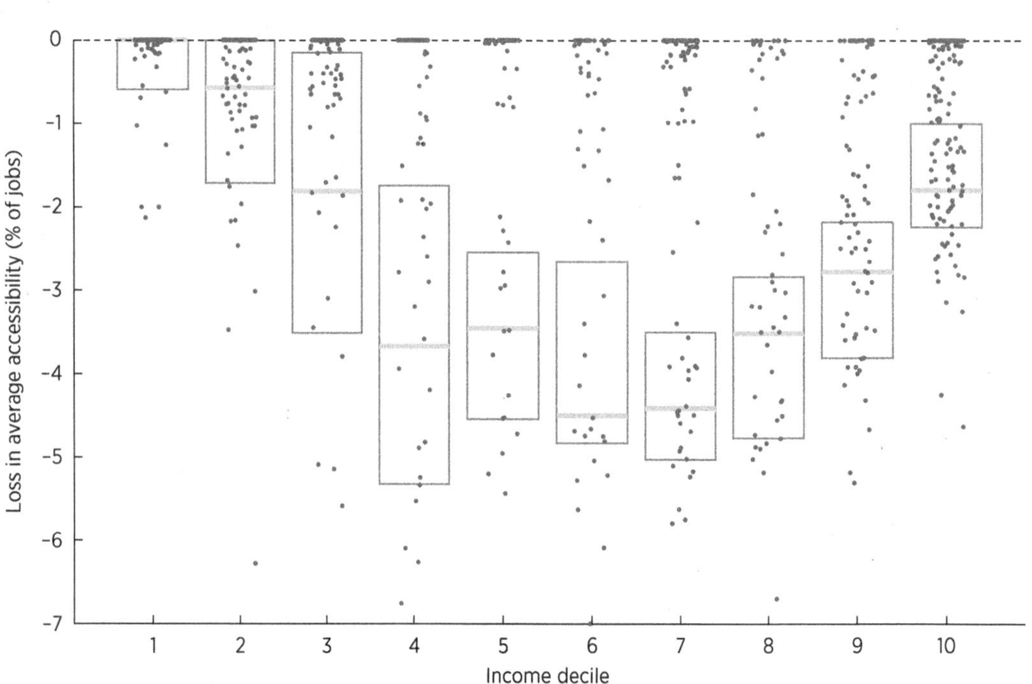

Source: Nell et al. 2023.

BOX O.2

Gender and climate mitigation policies in agriculture

Despite women's important contribution to agriculture, persisting gender inequalities mean they tend to have less access to resources, including land, inputs, financial services, education, and decent employment opportunities (Erman et al. 2021). Gender roles in agriculture and structural differences have implications not only for how climate mitigation policies affect social equity outcomes but also for how different groups adopt these policies and whether the policies facilitate a just transition for all. Mitigation policies and instruments that do not recognize gender dynamics and related transitional challenges could exacerbate existing inequalities. For example, a study in India shows that shifting rice production from conventional to direct-seeded or machine-transplanted methods could affect women more than men (Gartaula et al. 2020).

Making mitigation policies gender-responsive also poses a challenge because those who can benefit often lack political voice. Evidence shows this lack of power, voice, and recognition exists from household and local levels (Larson et al. 2015) to national and international decision-making and governance frameworks (Gautam et al. 2022). Recognizing women's knowledge of and role in sustainable practices and creating meaningful opportunities for them in decision-making can increase mitigation effectiveness and reduce gender gaps. Undertaking gender analysis in mitigation policies and programs—including financing mechanisms and technological development—can help ensure distributive and procedural justice.

Source: Kabir, De Vries Robbe, and Godinho, forthcoming.

Governments can use complementary policies to reduce distributional effects

Complementary policies can ensure that climate policies are affordable and do not have a negative effect on poor and vulnerable populations. For example, recycling just a fraction of carbon pricing revenues or repurposing subsidies through direct transfer can usually make reforms pro-poor. One study of Latin American countries, for example, estimates that recycling about 20 percent of savings from subsidy reforms would fully alleviate consumption impacts on the bottom 40 percent of the income distribution (Feng et al. 2018).

But compensating people is hard: it requires appropriate systems and delivery mechanisms, including broad, strong, and flexible social protection systems. The heterogeneity of impacts makes it difficult to target transfers to support the most affected and vulnerable households. In Chile, Colombia, Ecuador, and Uruguay, an estimated 3 to 4 percent of all households are not covered by social transfers, even though they are both among the poorest 20 percent and among the 20 percent most affected by a fuel price change (Missbach, Steckel, and Vogt-Schilb 2022). Spatial effects and income and employment transitions are even harder to manage and require active labor market policies and proactive sectoral and regional transition strategies and support. Importantly, loss of employment goes beyond a simple loss of income and can affect people's status, culture, and family or community life.

As well as fostering winners, green industrial policy can help reduce impacts for potential policy losers and smooth the transition (Cullenward and Victor 2020; Hallegatte, Fay, and Vogt-Schilb 2013). When behaviors are weakly responsive to prices, when firms and households lack substitution options to adjust to a change in fossil fuel prices, or when prices cannot be changed for political reasons, implementing green industrial policies first can transform the capital stock or create substitution options while minimizing short-term social costs (Rozenberg, Vogt-Schilb, and Hallegatte 2020). Rather than a substitute for other policies, such as carbon pricing, green industrial policies can therefore appear as a complementary, or preparatory step. And, when the main political economy obstacle is the political economy of concentrated impacts, governments can use green industrial policies to support sunset industries to facilitate their downscaling or adjustment. To reap the full benefits of green industrial policies, however, countries need to carefully manage some political economy risks, including corruption, policy capture, and distributional conflicts.

Some communities or regions specialize heavily in activities with high carbon intensity, such as coal mining, and need a place-based approach to ensure a just and acceptable transition. The experiences of European coal regions, which lost their coal-related revenues and employment decades ago, illustrates how well-managed coal transitions can minimize long-term negative effects. But managing such major economic transitions takes time, and governments need to consider effects on labor; social, human, and economic development; local ownership, participation, and mobilization; stakeholder inclusion; and inclusiveness. Rather than adopting simple compensation mechanisms that focus only on employment impacts, successful transitions have targeted social, human, and economic development interventions and have had strong local ownership and engagement. The Dutch 10-year coal phase-out, which included substantial support for workers who lost their jobs and was supported by the trade union, shows that a well-planned transition need not have severe long-term adverse impacts.

Widely used to support the transition of distressed communities, place-based policies can include a range of measures, from tax incentives and expenditures to manufacturing extension and training programs. Although it is generally preferable to invest in people rather than places (World Bank 2009), governments can justify place-based interventions that reduce barriers to—or the costs of—migration, increase spatial equity, or help affected regions fulfill their economic potential (Bartik 2020). The research on the costs and benefits of place-based policies suggests that results depend on scale and design (Grover, Lall, and Maloney 2022). But experience shows that these interventions need to include multiple instruments, such as transportation investments to improve connections, fiscal incentives and direct service provisions, and a package of measures to foster skills, enterprise development, and innovation. Considerable evidence demonstrates that tax incentives alone are not enough for a policy to succeed: one study across 77 countries finds that infrastructure and trade facilitation have a significant positive impact and that tax and other financial incentives are less important (Farole 2011).

Policy process: Use public engagement and communication to improve policies and their legitimacy

Engagement helps policy makers design good policies and levels the influence playing field, preventing capture by the most influential interest groups. Different groups can voice their priorities and values through civic engagement processes, giving policy makers crucial information related to objective setting, instrument choice, and framing. Civic engagement can bring different communities' concerns to the surface, allowing policy makers to identify where complementary measures are needed. Including a wider array of actors in policy processes can also spur policy innovation, allow governments to test reactions to policy options ahead of implementation, and boost process legitimacy (box O.3).

BOX O.3

Process legitimacy depends on how decisions are made

Process legitimacy—one of four critical dimensions of social sustainability alongside social cohesion, inclusion, and resilience—is about how policy making and implementation happen, how consistent they are with a given context, and how legitimate most actors perceive them to be. Specifically, it has to do with the extent to which actors in society accept who has authority, what goals are formulated, and how decisions are made and implemented.

Process legitimacy is strong when actors believe that decisions are made by credible authorities in ways that align with their values and reflect accepted rules and norms relating to decision-making, including measures that support conflict resolution and compromises. Transparent and participatory processes, and desirable or acceptable outcomes, can enhance process legitimacy. Including and engaging potential policy losers is particularly important when policies might incur costs.

Source: Barron et al. 2023.

FIGURE O.5. Share of population that reported being "satisfied" or "very satisfied" with El Salvador's 2011 subsidy reform, 2011–13

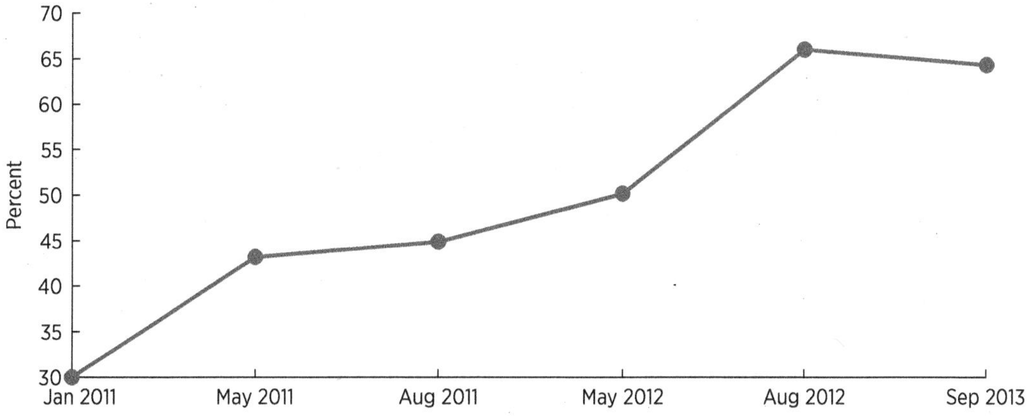

Source: Calvo-Gonzalez, Cunha, and Trezzi 2015.

Public perception can be an important driver of opposition to climate policies, even when policies follow a sound and progressive design. This is because, without public trust, even well-designed, well-intentioned promises of compensation and redistribution can lack credibility. Although El Salvador's 2011 gas subsidy reform benefited households in all but the top two deciles of the income distribution, it was unpopular, especially among the lower-income groups that would benefit most, in large part because of misinformation and mistrust of the government's ability to implement the policy. Perceptions improved gradually—and significantly—with the share of people expressing support for the policy increasing from 30 percent to 60 percent over a year and a half (figure O.5). Similarly, empirical evidence from Indonesia shows a direct link between opposition to fossil fuel subsidy reform and local perceptions of corruption (Kyle 2018). When corruption levels are perceived to be low, poor households are more than two-and-a-half times more likely to support than to oppose fuel subsidy reform.

Important tools include policy communication and awareness raising, which equip different actors with the information needed to advocate for or appropriately respond to policy measures. Transparency helps prevent and allay concerns about policy capture, and such trust is essential for process legitimacy. For example, when Nigeria reformed fossil fuel subsidies in 2012, its failure to explain the intended use of about half of the subsidy savings raised suspicions of corruption, further fueling antigovernment protests (Alleyne and Hussain 2013). Indonesia's successful 2005 subsidy reform, by contrast, included a wide-scale, well-prepared communication campaign that highlighted how savings would be recycled through a cash transfer to compensate for the impacts of reforms.

A way forward: Enacting climate action

For policy makers, introducing the policies needed to achieve climate objectives will be challenging and will require time, resources, and political capital. Unsurprisingly, governments tend to rely more on climate policies that use "pull" (subsidies and incentives) and nonmarket instruments because they tend to face less opposition than "push" (taxes and fines) and market instruments (Drews and van den Bergh 2016). The 2022 US Inflation Reduction Act, the world's largest fiscal package for climate mitigation to date, comprises only "pull" instruments, such as public investment and tax credits for clean energy, fuel and vehicles, conservation, and reducing air pollution.

But not all climate reforms will be popular or easy. Achieving long-term goals will mean changing and navigating the political economy. More disruptive or transformative policies—which can often also improve efficiency, capacity, and productivity and deliver benefits for development—require governments and policy makers to trigger shifts in actors' ideas, interests, or influence to generate enough support. This could mean establishing the right governance framework, including civic engagement platforms, to provide information that helps actors (re)form ideas about climate change and (re)consider their interests, and allows them to influence the climate policy agenda. At the strategic level, it also means selecting policies that not only are based on economic costs and benefits but also influence the political economy and build consensus and momentum toward shared objectives. At the policy level, it means trying to make more difficult policy instruments acceptable or even attractive for key stakeholders by engaging with those stakeholders early on, reflecting their interests in policy design, and clearly communicating with them.

The urgency of the climate change challenge means that we cannot afford to wait for the right governance and political economy context before starting to implement climate policies. Instead, institutional changes and capacity building will have to happen in parallel with the implementation of the best policies, chosen among those that are feasible and can create momentum and facilitate further action in the future. This process will have to be supported by a combination of complementary policies—to protect the most vulnerable and to prevent or compensate for concentrated impacts on sectors or regions—and strong communication and engagement processes to make sure policy design is informed by stakeholders and that stakeholders understand and accept policy choices. Only such an approach, built for and informed by the political economy, can unlock climate policy progress and create pathways to more rapid transformative change.

Notes

1. NewClimate Institute, Climate Policy Database, https://climatepolicydatabase.org.
2. Targets or strategies are considered "binding" when individuals and institutions in the public and private sectors must comply with them, for instance, because the targets or strategies are part of national legislation.
3. *Sunrise* sectors or industries are expected to grow in productivity and competitiveness over time and benefit from climate and industrial policies; *sunset* sectors or industries are those in decline.

References

Alleyne, T. S. C., and M. Hussain. 2013. *Energy Subsidy Reform in Sub-Saharan Africa: Experiences and Lessons.* Departmental Paper No. 2013/002. Washington, DC: International Monetary Fund. https://www.imf.org/en/Publications/Departmental-Papers-Policy-Papers/Issues/2016/12/31/Energy-Subsidy-Reform-in-Sub-Saharan-Africa-Experiences-and-Lessons-40480.

Barron, P., L. Cord, J. Cuesta, S. A. Espinoza, G. Larson, and M. Woolcock. 2023. *Social Sustainability in Development: Meeting the Challenges of the 21st Century.* Washington, DC: World Bank. https://doi.org/10.1596/978-1-4648-1946-9.

Bartik, T. J. 2020. "Using Place-Based Jobs Policies to Help Distressed Communities." *Journal of Economic Perspectives* 34 (3): 99–127. https://doi.org/10.1257/jep.34.3.99.

Calvo-Gonzales, O., B. Cunha, and R. Trezzi. 2015. "When Winners Feel Like Losers: Evidence from an Energy Subsidy Reform." Policy Research Working Paper 7265, World Bank, Washington, DC. http://hdl.handle.net/10986/21998.

Cullenward, D., and D. G. Victor. 2020. *Making Climate Policy Work.* New York: John Wiley and Sons.

Damania, R., E. Balseca, C. de Fontaubert, J. Gill, K. Kim, J. Rentschler, J. Russ, and E. Zaveri. 2023. *Detox Development: Repurposing Environmentally Harmful Subsidies.* Washington, DC: World Bank. http://hdl.handle.net/10986/39423.

Dorband, I. I., M. Jakob, M. Kalkuhl, and J. C. Steckel. 2019. "Poverty and Distributional Effects of Carbon Pricing in Low- and Middle-Income Countries—A Global Comparative Analysis." *World Development* 115: 246–57. https://doi.org/10.1016/j.worlddev.2018.11.015.

Dorband, I. I., M. Jakob, J. C. Steckel, and H. Ward. 2022. "Double Progressivity of Infrastructure Financing through Carbon Pricing—Insights from Nigeria." *World Development Sustainability* 1: 100011. https://doi.org/10.1016/j.wds.2022.100011.

Dorband, I. I. Forthcoming. "Distributional Effects of Climate and Development Policies across Households and Workers in Low- and Middle-Income Countries." Background paper for this report.

Douenne, T. 2020. "The Vertical and Horizontal Distributive Effects of Energy Taxes: A Case Study of a French Policy." *Energy Journal* 41 (3). https://doi.org/10.5547/01956574.41.3.tdou.

Drews, S., and J. C. J. M. van den Bergh. 2016. "What Explains Public Support for Climate Policies? A Review of Empirical and Experimental Studies." *Climate Policy* 16 (7): 855–76. https://doi.org/10.1080/14693062.2015.1058240.

Dubash, N. K. 2021. "Varieties of Climate Governance: The Emergence and Functioning of Climate Institutions." *Environmental Politics* 30 (1): 1–25. https://doi.org/10.1080/09644016.2021.1979775.

Dubash, N., A. V. Pillai, C. Flachsland, K. Harrison, K. Hochstetler, M. Lockwood, R. Macneil, M. Mildenberger, M. Paterson, F. Teng, and E. Tyler. 2021. "National Climate Institutions Complement Targets and Policies." *Science* 374 (6568): 690–93.

Erman, A., S. A. De Vries Robbe, S. Thies, K. Kabir, and M. Maruo. 2021. "Gender Dimensions of Disaster Risk and Resilience. Existing Evidence." World Bank, Washington, DC. http://hdl.handle.net/10986/35202.

Farole, T. 2011. *Special Economic Zones in Africa: Comparing Performance and Learning from Global Experience.* Directions in Development. Washington, DC: World Bank. http://hdl.handle.net/10986/2268.

Feindt, S., U. Kornek, J. M. Labeaga, T. Sterner, and H. Ward. 2021. "Understanding Regressivity: Challenges and Opportunities of European Carbon Pricing." *Energy Economics* 103: 105550. https://doi.org/10.1016/j.eneco.2021.105550.

Feng, K., K. Hubacek, Y. Liu, E. Marchán, and A. Vogt-Schilb. 2018. "Managing the Distributional Effects of Energy Taxes and Subsidy Removal in Latin America and the Caribbean." *Applied Energy* 225: 424–36. https://doi.org/10.1016/j.apenergy.2018.04.116.

Gartaula, H., T. B. Sapkota, A. Khatri-Chhetri, G. Prasad, and L. Badstue. 2020. "Gendered Impacts of Greenhouse Gas Mitigation Options for Rice Cultivation in India." *Climate Change* 163: 1045–63. https://doi.org/10.1007/s10584-020-02941-w.

Gautam, M., D. Laborde, A. Mamun, W. Martin, V. Pineiro, and R. Vos, R. 2022. "Repurposing Agricultural Policies and Support: Options to Transform Agriculture and Food Systems to Better Serve the Health of People, Economies and the Planet." World Bank, Washington, DC. http://hdl.handle.net/10986/36875.

Godinho, C. 2022. "What Do We Know about the Employment Impacts of Climate Policies? A Review of the Ex Post Literature." *WIREs Climate Change* 13 (6). https://doi.org/10.1002/wcc.794.

Godinho, C., S. Hallegatte, and J. Rentschler. Forthcoming. "What Do We Know about the Political Economy of Climate Policies? Reviewing the Literature through the 4 i's—Institutions, Interests, Ideas, and Influence." Background paper for this report.

Greve, H., and J. Lay. 2023. "'Stepping Down the Ladder': The Impacts of Fossil Fuel Subsidy Removal in a Developing Country." *Journal of the Association of Environmental and Resource Economists* 10 (1): 121–58. https://doi.org/10.1086/721375.

Grover, A. G., S. V. Lall, and W. F. Maloney. 2022. *Place, Productivity and Prosperity: Revisiting Spatially Targeted Policies for Regional Development—Overview (Vol. 2)*. Washington, DC: World Bank. http://documents.worldbank.org/curated/en/099130001182297916/P1725410e343500aa0a3de030d5f1599b46.

Hale, T. 2020. "Catalytic Cooperation." *Global Environmental Politics* 20 (4): 73–98.

Hallegatte, S., M. Fay, and A. Vogt-Schilb, A. 2013. "Green Industrial Policies: When and How." Policy Research Working Paper 6677, World Bank, Washington, DC.

Kabir, K., S. A. De Vries Robbe, and C. Godinho. forthcoming. "Climate Mitigation Policies in Agriculture: A Review of Socio-Political Barriers." Background paper for this report.

Keohane, R. O., and M. Oppenheimer. 2016. "Paris: Beyond the Climate Dead End through Pledge and Review?" *Politics and Governance* 4 (3): 142–51.

Kyle, J. 2018. "Local Corruption and Popular Support for Fuel Subsidy Reform in Indonesia." *Comparative Political Studies* 51 (11). https://doi.org/10.1177/0010414018758755.

Larson, A. M., T. Dokken, A. E. Duchelle, S. Atmadja, I. A. P. Resosudarmo, P. Cronkleton, M. Cromberg, W. Sunderlin, A. Awono, and G. Selaya. 2015. "The Role of Women in Early REDD+ Implementation: Lessons for Future Engagement." *International Forestry Review* 17 (1): 43–65.

Li, L., and A. Taeihagh. 2020. "An In-Depth Analysis of the Evolution of the Policy Mix for the Sustainable Energy Transition in China from 1981 to 2020." *Applied Energy* 263: 114611.

Mealy, P., M. Ganslmeier, C. Godhino, and S. Hallegatte. Forthcoming. "Climate Policy Feasibility Frontiers: A Tool for Realistic and Strategic Climate Policy Making." Background paper for this report.

Metcalf, G. E., and J. H. Stock. 2020. "Measuring the Macroeconomic Impact of Carbon Taxes." *AEA Papers and Proceedings* 110: 101–106. https://doi.org/10.1257/pandp.20201081.

Missbach, L., J. C. Steckel, and A. Vogt-Schilb. 2022. "Cash Transfers in the Context of Carbon Pricing Reforms in Latin America and the Caribbean." IDB Working Paper Series, Inter-American Development Bank, Washington, DC. http://dx.doi.org/10.18235/0004568.

Nell, A., D. Herszenhut, C. Knudsen, S. Nakamura, M. Saraiva, and P. Avner. 2023. "Carbon Pricing and Transit Accessibility to Jobs: Impacts on Inequality in Rio de Janeiro and Kinshasa." Policy Research Working Paper 10341, World Bank, Washington, DC. https://doi.org/10.1596/1813-9450-10341.

NewClimate Institute. 2022. *Climate Action Tracker: Warming Projections Global Update (November)*. Cologne: NewClimate Institute. https://climateactiontracker.org/documents/1094/CAT_2022-11-10_GlobalUpdate_COP27.pdf.

Peng, W., G. Iyer, V. Bosetti, V. Chaturvedi, J. Edmonds, A. A. Fawcett, S. Hallegatte, D. G. Victor, D. van Vuuren, and J. Weyant. 2021. "Climate Policy Models Need to Get Real about People—Here's How." *Nature* 594: 174–76. https://doi.org/10.1038/d41586-021-01500-2.

Persson, T., and G. Tabellini. 2002. "Political Economics and Public Finance." In *Handbook of Public Economics*," Volume 3, edited by A. J. Auerbach and M. Feldstein, 1549–659. Elsevier. https://doi .org/10.1016/S1573-4420(02)80028-3.

Pillai, V. A., and N. K. Dubash. 2021. "The Limits of Opportunism: The Uneven Emergence of Climate Institutions in India." *Environmental Politics* 30 (supl): 93–117. https://doi.org/10.1080/09644016 .2021.1933800.

Rozenberg, J., A. Vogt-Schilb, and S. Hallegatte. 2020. "Instrument Choice and Stranded Assets in the Transition to Clean Capital." *Journal of Environmental Economics and Management* 100 (2020): 102183.

Sharpe, S., and T. M. Lenton. 2021. "Upward-Scaling Tipping Cascades to Meet Climate Goals: Plausible Grounds for Hope." *Climate Policy* 21 (4): 421–33. https://doi.org/10.1080/14693062 .2020.1870097.

UNFCCC (United Nations Framework Convention on Climate Change). 2023. "Technical Dialogue of the First Global Stocktake. Synthesis Report by the Co-facilitators on the Technical Dialogue." Report prepared for the UN Climate Change Conference, United Arab Emirates November/ December 2023. https://unfccc.int/documents/631600.

World Bank. 2009. *World Development Report 2009: Reshaping Economic Geography.* Washington, DC: World Bank. https://openknowledge.worldbank.org/entities/publication/58557d74-baf0-5f97 -a255-00482909810a.

World Bank. 2023. *Reality Check: Lessons from 25 Policies Advancing a Low-Carbon Future*. Climate Change and Development Series. Washington, DC: World Bank. http://hdl.handle.net/10986/40262.

World Bank Group. 2022. *Climate and Development: An Agenda for Action—Emerging Insights from World Bank Group 2021–22 Country Climate and Development Reports*. Washington, DC: World Bank. http://hdl.handle.net/10986/38220.

1 Political Economy

A Major Barrier to Aligning Climate Policies with Commitments

In 2015, 195 governments agreed to strengthen the global response to the urgent threat of climate change by setting clear climate goals in the landmark Paris Agreement. They agreed to hold global warming at well below 2°C above preindustrial levels and pursue efforts to limit it to 1.5°C. The subsequent 2021 Glasgow Pact affirms the need to reduce global carbon dioxide (CO_2) emissions to net zero by midcentury to limit warming to 1.5°C. *Net zero* means reducing emissions as close to zero as possible and compensating for the remainder with carbon removals through natural carbon sinks and technological solutions. Staying within these limits should stabilize the climate and keep adaptation needs within manageable bounds, safeguarding sustainable development and poverty eradication efforts.

As of March 2023, 172 countries had submitted a new or updated nationally determined contribution in line with the Paris Agreement's ratcheting mechanism to better align country commitments with global goals (see box 1.1 later in this chapter). In addition, more than 70 countries covering 76 percent of global emissions have pledged to reach net zero, including most major polluters, such as China, the European Union, and the United States.

On the whole, countries' emissions reduction commitments are converging on Paris Agreement targets. According to the Climate Action Tracker, an emissions pathway based on nationally determined contribution targets alone would likely lead to warming of 2.4°C in 2100; when factoring in countries' pledges, projections decrease to the Paris Agreement's upper global warming limit of 2°C (NewClimate Institute 2022). Although they will need to ratchet their emissions reduction commitments further to align them with the 1.5°C limit, commitments are slowly converging toward global goals, which is a major success for the Paris Agreement and its bottom-up approach.

The urgent need to align policies with pledges and commitments

Despite increasingly ambitious high-level commitments, most countries are not on track to achieve their own targets or to contribute their fair share to reducing global emissions.

In fact, following decades of growth, CO_2 emissions surpassed prepandemic levels in 2022, reaching an all-time high of more than double the emission levels of 50 years ago (UNFCCC 2023).

To align commitments with action, all countries urgently need to implement policies that will rapidly decarbonize their economies and development paths. To do so, they will need to develop and implement policies to decarbonize the electricity supply; deploy efficiency measures, fuel substitution, and electrification to reduce emissions from transportation, buildings, and industry; advance low-carbon agricultural practices; protect and expand forests and other natural carbon sinks; and address emissions related to lifestyle, behavior, and consumption with demand-side measures (Fay et al. 2015).

The political economy as a major barrier to more ambitious climate policy

Despite long-harbored concerns about their economic impacts, these climate policies can work for the economy. If well designed and implemented, they can deliver benefits and contribute to economic development goals. The Intergovernmental Panel on Climate Change and much of the literature suggest that limiting global warming to 2°C or less would have a small, and possibly negative, aggregate economic cost (IPCC 2023). The World Bank Group's Country Climate and Development Reports show that, with early, well-designed actions, it is possible to reduce greenhouse gas (GHG) emissions by 70 percent by 2050 without compromising economic growth. Countries could also see immediate economic, productivity, and health benefits because efficiency, electrification, and the shift to renewable energy will improve air quality, reduce congestion, and reduce fuel spending and imports.

These analyses raise the following key question: Why are the opportunities to capture synergies between development, growth, poverty reduction, and climate objectives not explored more systematically? This book claims that a key reason is that such policies do not always work for politics: powerful vested interests, winners and losers, ideological battles, and institutional inertia are just some of the political economy issues that make it difficult to capture synergies between development and climate objectives. Unequal distributional impacts are undeniably part of the political economy challenge, but political economy challenges go beyond distributional impacts. In recent years, several ambitious climate-related policies have triggered intense public backlash and become highly political issues, with many factors affecting public and political acceptability. There are also governance and institutional challenges to consider. It is these political economy constraints, rather than technical limits, that pose the main barrier to more, and more stringent, climate policies.

Widespread progress in spite of political economy barriers

These political economy barriers are not unmovable, and countries have not been idle: they have implemented climate policies that are limiting emissions growth. The number and coverage of climate policies have expanded considerably over the past three decades, and evidence shows that they have curbed and continued to curb emissions growth (Eskander and Fankhauser 2020; IPCC 2022). According to the NewClimate Institute's Climate Policy Database, countries have announced more than 4,500 climate policies over the last three decades (figure 1.1).[1] Key economywide policies, such as ending fossil fuel subsidies and carbon pricing, are progressing but remain less frequent than sectoral policies. Long-term targets and planning help countries coordinate across sectors, with short-term milestones informing sector-level polices and actions and providing predictability and signals to stakeholders.

The companion report to this book, *Reality Check: Lessons from 25 Policies Advancing a Low-Carbon Future* (World Bank 2023a), provides examples of successful implementation of climate policies, even in difficult political economy contexts. High-visibility failures, or unrest as recently seen in Ecuador and France, hide the large and growing number of climate policies that are being successfully implemented. The *Reality Check* report finds that public and political support, strong institutions, cross-party backing, broad engagement, and flexibility in design have played key roles in these successes. Collaboration among diverse stakeholders, including government, private sector, civil society, and academics, is vital. Capacity building aids policy design and execution, with contextual adaptations necessary for success.

Many of the examples are not first-best policies or even best practice: to achieve successful implementation, governments often had to compromise. They faced institutional capacity constraints and had to manage trade-offs with other policy objectives. Some interventions were just the first step, and most countries are adjusting policy design as they draw lessons from real-world implementation. But these interventions managed to draw enough support to be implemented, and to create momentum toward more climate action. Examples include the following:

FIGURE 1.1. Climate policies announced globally, 1980–2020

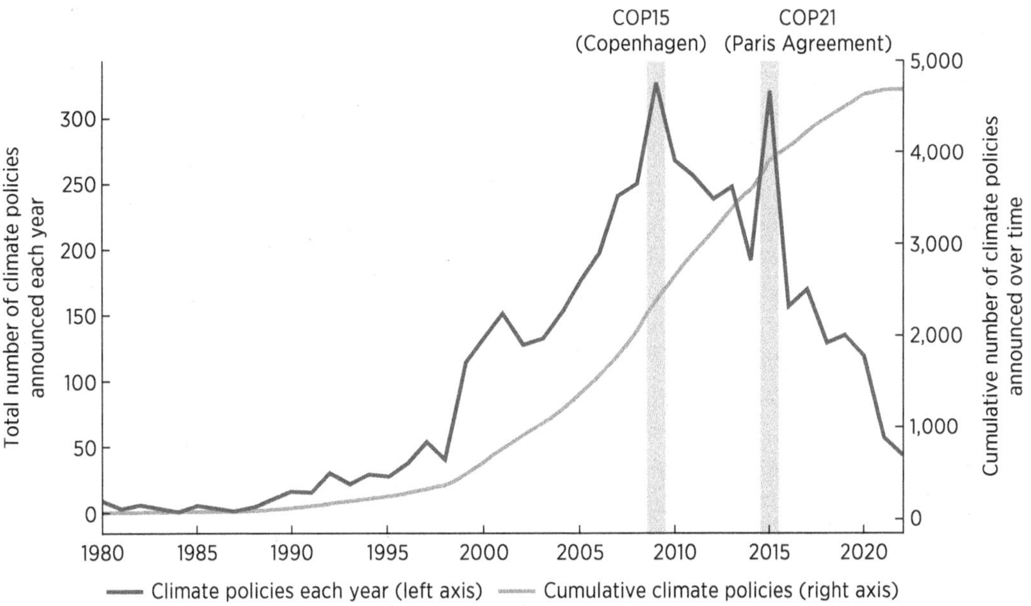

Source: World Bank 2023a, based on data from NewClimate Institute, Climate Policy Database (https://climatepolicydatabase.org/).
Note: COP15 = 15th session of the Conference of the Parties to the 1992 United Nations Framework Convention on Climate Change (2009); COP21 = 21st session of the Conference of the Parties to the 1992 United Nations Framework Convention on Climate Change (2015).

- *Costa Rica's National Decarbonization Plan.* The government made a concerted effort to convince diverse stakeholders of the need for bold action, some of whom did not originally share that ambition. The National Decarbonization Plan is one of the most ambitious global strategies for low-carbon development for a middle-income country. Net benefits of implementing the new plan are estimated at US$41 billion through 2050, which helped Costa Rica mobilize at least US$2.4 billion in international concessional finance.
- *The Arab Republic of Egypt's energy subsidy reform.* Egypt offers an example of successful fossil fuel subsidy reform—a notoriously challenging policy to implement. Energy subsidy reform has eased fiscal pressures, with the budget deficit falling from 12.9 percent to 8.1 percent of gross domestic product between 2013 and 2019. It has also encouraged greater private investment in clean energy, with solar and wind generation growing almost threefold between 2014 and 2019. Proactive communications and a boost in social protection mechanisms played strong roles in facilitating the reforms.
- *British Columbia's carbon tax.* Introduced by the Canadian province in 2008, the tax covers 70 percent of GHG emissions. Strong communication efforts facilitated implementation, with extensive empirical evidence finding that the tax reduced emissions and inequality and raised growth and employment. The reform now receives majority support from citizens.
- *Climate-smart agriculture in Africa's Sahel region.* Thanks to targeted interventions, farmers have adopted low-cost, efficient traditional practices, such as agroforestry and conventional rainwater harvesting techniques, to capture rainfall, reduce runoff, restore soils, and improve agricultural productivity. In Niger, farmer-managed natural regeneration increased yields by 16–30 percent between 2003 and 2008, while adding nearly 5 million hectares of tree cover.
- *Colombia's mandatory green building code.* Along with this code, enacted in 2015, the government introduced tax incentives for technical solutions such as insulation and energy-efficient air conditioning systems. By the end of 2022, 11.5 million square meters of green space had been built or were under construction (World Bank 2023c). Between 2021 and 2022, 27 percent of new buildings were certified as green under the International Finance Corporation's EDGE program.
- *India's national solar mission.* The support of federal, state, and local policies and regulations has contributed to the country's solar success. One of the world's most rapidly growing solar markets, India saw solar go from 4 percent to 13 percent of power generation between 2014 and 2022. The private sector has been heavily involved in creating India's renewable energy market, investing US$130 billion since 2004. The country is becoming a domestic manufacturing hub for solar panels, which has created new green jobs.
- *Power sector reforms in Kenya.* These reforms have made Kenya one of the most successful countries in attracting private financing for clean power assets, and an investment destination for independent power producers. Because of successful energy sector reforms, as much as 70 percent of new power generation capacity in the country is renewable energy, including geothermal, solar, and wind. Since 2000, the CO_2 intensity of power generation has fallen fourfold. As well as reducing GHGs, these reforms have had substantial development benefits, making electricity supply more reliable and increasing people's access to energy.

The scope of this book

This book aims to understand why climate policies have been successfully implemented in some places but have triggered strong opposition elsewhere, and to provide a framework to help policy makers reproduce these successes and navigate political economy constraints. The many political economy analysis tools already developed, inside and outside of the World Bank (Fritz, Levy, and Ort 2014; Hudson and Leftwich 2014; World Bank 2017), do not provide a discussion of climate change challenges. This book contributes to this line of work by focusing on low-carbon development and countries' emissions reduction objectives. It does not, however, cover the whole range of issues relevant for the political economy of climate change.

First, this book focuses on low-carbon development and commitments related to GHG emissions but does not cover similar political economy challenges faced by adaptation and resilience goals. As discussed in *World Development Report 2014: Risk and Opportunity—Managing Risk for Development* (World Bank 2013), there are similarly strong political economy barriers to implementing resilience and adaptation policies (for example, when flood management has an impact on land values and therefore the redistribution of wealth). Strong vested interests and political conflicts will also affect the feasibility and viability of policies related to sharing resources, such as water across users, including for agriculture and irrigation, hydropower generation, health and well-being, and maintenance of healthy ecosystems. Although these issues are important, they are not the topic of this book.

Second, this book focuses on national-level political economy challenges, without going deeply into the important international dimensions. For example, we do not discuss how to manage historical responsibility for today's GHG concentrations and the unequal use of the carbon budget available for keeping global warming well below 2°C. Historical responsibility and the concept of Common but Differentiated Responsibilities of the United Nations Framework Convention on Climate Change and the Paris Agreement are essential dimensions of the discussion regarding the required ambition of countries' targets and policies, the needs for concessional and nonconcessional financing, and how to help poorer countries and people manage the unavoidable "loss and damage" caused by climate change. These questions are related, because political feasibility at the country level depends on the fairness of global processes and the fair contribution of all countries; however, this book does not focus on such international challenges.

Third, this book focuses on how governments with a genuine commitment to their stated climate goals can achieve those goals, despite facing a range of other objectives and a complex political economy context. Governments have different incentives and priorities when it comes to climate action. These are affected and constrained by the domestic context, as discussed later in this book, and by the international architecture, including the Paris Agreement (box 1.1) and other agreements and relationships. Although a gap exists between the objectives stated in the international arena and the priorities discussed in the domestic context, World Bank (2023a) provides ample evidence that countries are working toward their stated goals. This book focuses on the barriers to increasingly rapid actions—barriers created by the political economy context.

This book is part of a broader work program. Further work on climate governance, institutions, and political economy is under way at the World Bank to support an increasing range of country analytics, engagements, and lending operations. That work aims to contribute to and support effective policy actions, as well as enable public and private investments.

BOX 1.1

The Paris Agreement ratcheting mechanism

The 2015 Paris Agreement is the second operational agreement reached under the United Nations Framework Convention on Climate Change. The first, the 1997 Kyoto Protocol, set emissions reduction targets for industrialized countries. Attempts to build consensus on a follow-up agreement to establish legally binding targets for all major economies failed at the 15th session of the Conference of the Parties to the 1992 United Nations Framework Convention on Climate Change (COP15) in 2009. Countries took an alternative approach, built around voluntary, nationally determined contributions, or NDCs (which are not legally binding). Instead of top-down targets or enforcement mechanisms, the resulting Paris Agreement relies on an "ambition ratcheting" mechanism, whereby countries regularly take stock of progress and increase the stringency of their NDCs accordingly (figure B1.1.1).

Some authors question the Paris Agreement's ability to achieve its global objectives because it lacks mechanisms to solve the free rider problem and to enforce the agreement (Bang, Hovi, and Skodvin 2016; Barrett and Dannenberg 2016; Keohane and Oppenheimer 2016; Nordhaus 2021; Sachs 2019). Hale (2020), however, makes the case that free riding is not the main obstacle to climate action. This is because the free riding (or prisoners' dilemma) framing neither captures the full complexity of the decarbonization challenge nor provides consistent explanation for the political behavior and outcomes observed in the past 25 years (Aklin and Mildenberger 2020). Hale identifies three essential characteristics of the climate change mitigation challenge that a traditional (repeated) prisoners' dilemma does not capture:

1. *Development-climate synergies (or joint products)*. Contributions to global public goods also yield "private" benefits, in the form of higher productivity and efficiency, reduced air pollution or congestion, and reduced energy costs and imports, as illustrated in the Country Climate and Development Reports (see World Bank Group 2022).
2. *Heterogeneity*. Countries face different costs and barriers to reducing greenhouse gas emissions, with some benefiting from higher renewable energy potential and others losing out because of reduced fossil fuel exports. Preferences also vary, with some countries exhibiting higher willingness to pay to protect the global climate.

FIGURE B1.1.1. **Paris Agreement: Mechanisms to ratchet ambition and monitor progress**

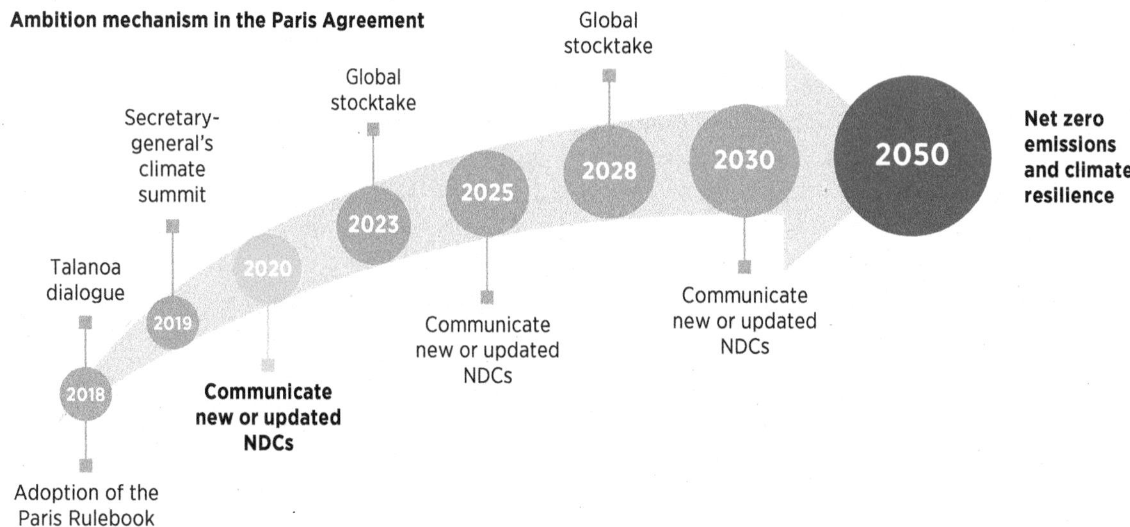

Source: Fransen et al. 2017.
Note: NDC = nationally determined contribution.

(Continued)

BOX 1.1

The Paris Agreement ratcheting mechanism (continued)

3. *Increasing returns*. Emissions reduction costs are not constant over time, and they depend on policies and investments, as illustrated by the declines in the cost of renewable energy, electricity storage, or electric mobility.

These characteristics mean that a group of early movers—those with lower emissions reduction costs, higher synergies with development, and more willingness to act—can have a transformational global impact by reducing costs for others. Thus, cooperative action becomes progressively self-reinforcing over time and helps overcome the current "lock-in" to carbon-intense patterns through innovation and scale (Bernstein and Hoffmann 2018; Farmer et al. 2019; see also chapter 3 of this book).

In this context, Hale (2020) claims that the Paris Agreement's main contribution is its support for "catalytic cooperation." It helps build action over time by changing preferences and lowering the cost of reducing emissions through technological change and shared experience on what works. The key weakness of a "pledge and review" approach is that it captures only what countries will commit to. In a "one-off" commitment system, this weakness would prevent the agreement's success. But, with evolving technologies, changing preferences, and increasing returns, countries may be willing to increase their own ambition over time. Through the NDC updates every five years, the Paris Agreement aims to create a dynamic feedback loop, capturing the impact of changed costs and preferences and translating them into enhanced ambition in the way ahead. It is well known that cooperation is easier when commitments are repeated continuously (Macy 1991), and most international institutions build on such an approach—for example, in trade regulations and tariffs, or commons such as the ozone layer.

Other authors have emphasized the impact of the review process to identify good practices, better anticipate costs and benefits, and increase the willingness to act (Abbott 2017; Aldy 2018; Chayes and Chayes 1995; Sabel and Victor 2017; Victor, Raustiala, and Skolnikoff 1998). The Paris Agreement includes several mechanisms through which exchange across countries can accelerate action—the NDC implementation review (article 13), the global stocktake (article 14), and the mechanism to "facilitate implementation of and promote compliance with the provision of the Paris Agreement" (in a "nonpunitive" manner) (article 15)—which are expected to facilitate countries' actions and reduce risks and costs over time.

Policy makers face different incentives in the international and domestic arenas, and pledges to the international community may differ from their domestic discourses and commitments (see box 1.3). Some scholars suggest that countries often achieve their international objectives only because those objectives require countries to do little more than they would in the absence of an international agreement (Downs, Rocke, and Barsoom 1996), and some highly visible cases of noncompliance suggest that international commitments can be empty promises, disconnected from domestic objectives. Other evidence, however, shows that goal setting can create domestic incentives and policies—for example, through benchmarking, international comparison, and "naming and shaming" (Biermann, Kanie, and Kim 2017; Kanie et al. 2017; Kelley 2017). Although the Paris Agreement does not include an assessment of NDCs or their implementation, its transparency framework enables third parties to compare countries' commitments and policies (NewClimate Institute and Climate Analytics 2019; van Asselt 2016).

International agreement also influences domestic political economy, affecting preferences and creating new constituencies. International climate governance has built a unique network of stakeholders and participants that goes beyond national governments to include nongovernmental organizations and community representatives, the academic world and think tanks, subnational actors such as cities and regions, and private sector actors, including industry associations and institutional investors (Betsill et al. 2015). The influence of this broader set of actors makes it more likely that governments will find a supportive constituency for strong climate policies (Bromley-Trujillo et al. 2016; Cao and Ward 2017; Stokes 2020; Urpelainen 2009).

The Paris Agreement is only one of the international agreements relevant for climate action. The broader international agenda—which includes the Sustainable Development Goals and the Addis Ababa Action

(Continued)

BOX 1.1

The Paris Agreement ratcheting mechanism (continued)

FIGURE B1.1.2. **Net GHG emissions per capita, by country income group, population, and GDP per capita, 2019**

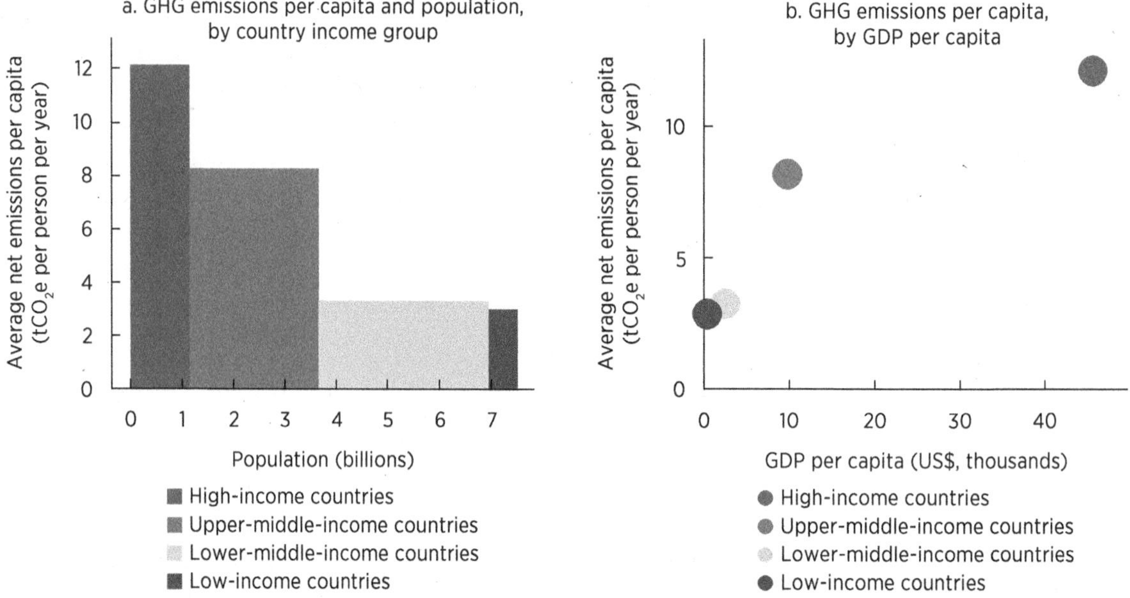

Sources: World Bank staff calculations based on data from Climate Watch (https://www.climatewatchdata.org) for net GHG emissions data and the World Bank DataBank for GDP and population data.
Note: GDP = gross domestic product; GHG = greenhouse gas; tCO_2e = tonnes of carbon dioxide equivalent.

Agenda—is also closely connected. As climate action becomes better mainstreamed and climate policies increasingly affect other domains, other agendas also become important. For example, the EU Carbon Border Adjustment Mechanism, the Inflation Reduction Act and its local content requirements, and the development of global green value chains all illustrate how climate policies increasingly connect to global, multilateral, and unilateral trade agreements. These parallel negotiation arenas create new incentives for countries to act on climate change—for example, to benefit from easier access to global consumer markets. Building on these opportunities could create additional incentives for action—for example, through climate clubs that would provide benefit for members that achieve certain levels of ambition or implement certain policies (Nordhaus 2021).

Nevertheless, the Paris Agreement faces difficult challenges, including on equity across countries. High-income countries have much higher emissions per capita than the rest of the world, and these countries have consumed a disproportional share of the carbon budget available to keep global warming well below 2°C (figure B1.1.2). And, despite declining emissions, their policies are still considered inconsistent with their own stated climate objectives (IPCC 2023; UNFCCC 2023). Regardless of whether the international community and network of international agreements create appropriate incentives and prevent free riding, there is a consensus that they do not provide appropriate resources for lower-income countries to tackle climate change. The discussion on the "means of implementation" remains extremely controversial. Most visible is high-income countries' failure to deliver on their commitment—made in Copenhagen in 2009 and included in the Paris Agreement in 2015—to a collective goal to "mobilize US$100 billion per year for the needs of developing countries in the context of meaningful mitigation action" (UNFCCC 2023). Understandably, many low- and middle-income country governments have asked for accelerated action in, and increased financial and technological support from, high-income countries before they implement more ambitious policies.

Building on this book and previous work focusing on public financial management and climate change, future work will focus on the full range of institutional capabilities that countries need to address climate change challenges, in key sectors such as energy, agriculture, and forestry, and for disaster risk management.

Outline of this book

Chapter 2 discusses climate governance, and how countries can adapt their institutional architecture to enable climate policies. It begins with an overview of climate governance, its forms, and its functions, and considers how countries can use it to enable more transformative action by reducing political economy barriers and increasing support over time. The chapter then focuses on national climate legislation, long-term strategies, and just transition frameworks—all of which can play a crucial role in building a public mandate for action and the institutional architecture to achieve it.

Chapter 3 explores how to prioritize and sequence climate policies to create ambitious policy pathways. The objective is to consider policy choices and prioritization not only in a static framework but also in terms of ongoing feedback dynamics between policies and political economy constraints. The chapter discusses how to select policies that align with a country's current capacity and institutional context, build future capacity, and influence the political economy over the long term. It then uses the Climate Policy Database to develop the Climate Policy Feasibility Frontier, a tool to help identify—in a given country context—the most realistic policies that can build momentum toward long-term objectives and transformation.

Chapter 4 focuses on policy design, or how to navigate political economy constraints and minimize negative distributional impacts. Distributional impacts cause opposition to climate policies from those who expect to experience negative impacts as well as from those concerned about the fairness of policy outcomes. The chapter explores distributional impacts along multiple dimensions, including income groups, sectors, regions, smaller spatial areas, and other preexisting dimensions of exclusion. It suggests approaches and processes for designing policies and their complementary actions in ways that improve design, monitor results, identify unintended consequences early, and build the needed legitimacy of climate action.

Finally, chapter 5 discusses public engagement and communication as tools in the policy process to maximize acceptability and legitimacy of reforms. It looks at different approaches to public engagement that can be applied throughout the policy development and implementation process to help policy makers navigate issues of fairness and equity, as well as limit undue influence and capture by vested interests. Acknowledging the fact that even well-designed reforms do not necessarily spur public support, the chapter discusses how well-targeted communication strategies can increase societal acceptance.

The 4i Framework for understanding political economy barriers

At first glance, the political economy barriers that impede climate change policies can appear so wide and varied that they defy any systematic assessment, let alone strategy. Governments manage many competing demands with constrained resources, and future climate change impacts are less visible and salient than immediate transition costs. Short-term political mandates can undermine action on long-term objectives; vested interests, misinformation, and lobbying can distort the policy discourse and public opinion;

high-level climate ambitions are not always in tune with the day-to-day priorities of political leaders and communities; low institutional capacity and poor governance can hamper progress; and repeated crises can change short-term priorities overnight.

Political economy has become a buzzword in climate discussions, especially when it comes to explaining climate policy failures. The following economy analysis tools can help governments build a comprehensive map of the political economy to inform their governance and policy strategies (World Bank 2017):

- *Institutional analysis* examines the formal and informal rules, norms, and practices that govern economic and political behavior, such as property rights, regulatory frameworks, and corruption. The World Bank's Climate Change Institutional Assessment is an application of this approach to climate change (see chapter 2).
- *Stakeholder analysis* identifies and explores the preferences and power of different actors, such as government officials, private sector actors, civil society groups, and international organizations.
- *Power analysis* analyzes the distribution and sources of power among actors in society, including how they exercise and contest it.
- *Narrative analysis* maps out predominant narratives, themes, and discourses to understand how actors construct meaning and understand political and economic processes, decisions, and outcomes.

A background paper for this book proposes a political economy framework—the 4i Framework—to provide a common and unifying ground from which to address the challenge (Godinho, Hallegatte, and Rentschler, forthcoming). As shown in table 1.1 and figure 1.2, the 4i Framework depicts the political economy as a system made up of mutually constitutive, interdependent, and dynamic relationships between the core elements—institutions, interests, ideas and influence—or the four i's (Jackson 2009).

Institutions matter

Governance and political institutions influence governments' ability to design and implement climate policies. The nature of the institutions influences their ability to implement these policies, but this link is complex and depends on many other factors.

TABLE 1.1. The four i's of political economy

Four i's	Definition	Examples
Institutions	The formal and informal rules, norms, and organizations that provide incentives and constraints for economic, political, and social behavior in society	Legislation and policy, legal systems, government agencies, tradition, norms, trust, and cultural practices
Interests	The wants, needs, and objectives that shape the preferences and behavior of actors	Income or profit (material); voting, representation, or political office (political); and belonging or status (social)
Ideas	The beliefs, values, and worldviews that shape the preferences and behavior of actors	Ideologies, identity, morals, cultural narratives, and scientific paradigms
Influence	The power, authority, and leverage that actors use to advance their interests and ideas and their interaction with each other and with institutions	Voting, civic organization, protest, media and communication, political lobbying, and bribery

Source: Godinho, Hallegatte, and Rentschler, forthcoming.

FIGURE 1.2. **The 4i Framework and four-pronged approach for climate policy**

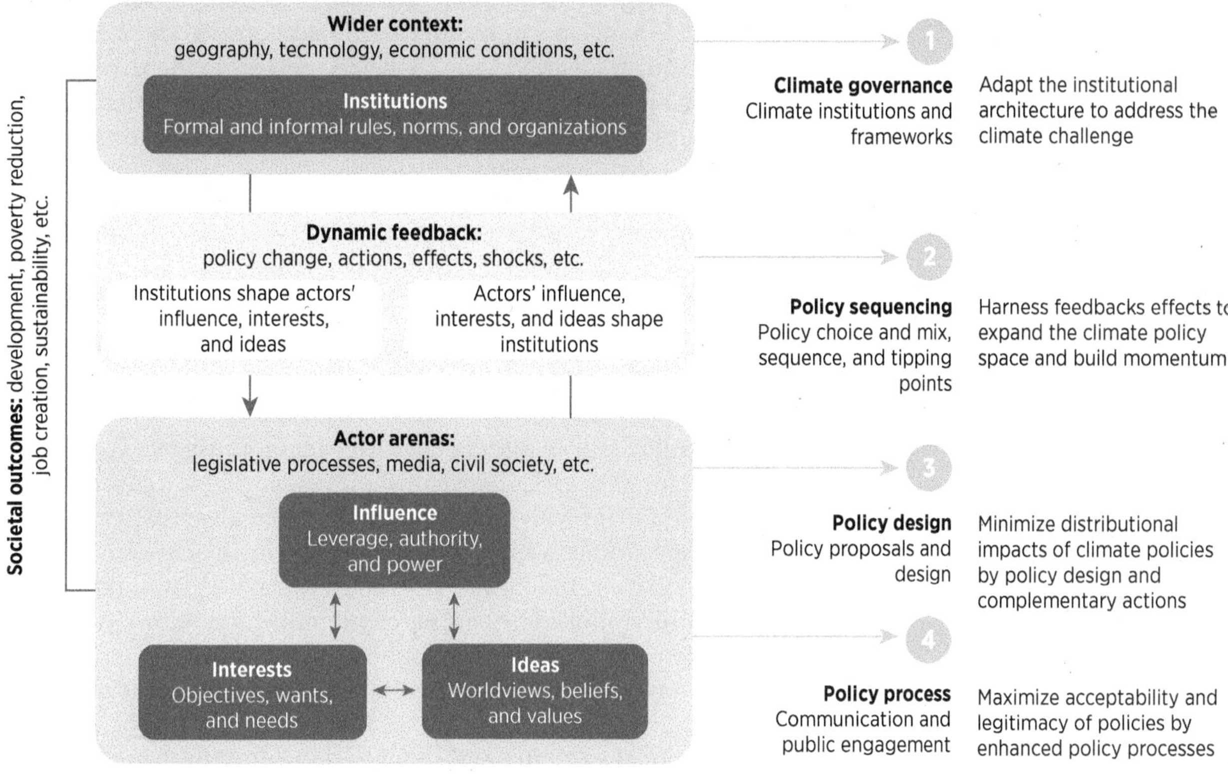

Source: Based on Godinho, Hallegatte, and Rentschler, forthcoming.

Despite important exceptions, on average, parliamentary democracies are more likely to implement climate policies and generally have lower emissions growth (Lachapelle and Paterson 2013; Lamb and Minx 2020). Proportional electoral systems seem to reduce the political costs of long-term climate policies, allowing governments to impose short-term costs on voters (Finnegan 2022). Countries with longer experience of democracy also appear to have more climate policies (Fredriksson and Neumayer 2013), possibly incentivized by democratic cultural norms such as accountability, a free press, and a stronger civil society. However, democracies are less likely to pass climate legislation ahead of elections (Lamb and Minx 2020), and having multiple veto points can make passing new laws or policies more difficult (von Stein 2022). Less democratic governments with strong state capacity and lower exposure to corruption may be able to implement swifter, more wide-ranging, and less popular climate policies (Beeson 2016; Michaelowa 2021). Political culture and public acceptability remain important in all contexts. Chapter 2 of this book focuses on how the political economy shapes the emergence of climate institutions and how governments can establish more strategic and sustainable climate governance.

The degree of (de)centralization in a country can also affect administrative capacity and effectiveness. On the one hand, environmental federalists have long argued that more centralized governance is needed for a coherent response to nationwide or global problems, such as climate change (Shobe 2020). Proponents of decentralized responses, on the other hand, argue that decentralization has two major political economy benefits. First, it allows for more ambitious subnational-level action in the absence of central government leadership (Steurer and Clar 2015). Second, it may make it easier for policy design and implementation to reflect local ideas and interests, increasing support for such policies. Looking at how India's fiscal federalism can be adjusted to account for climate change, Martinez-Vazquez and Zahir (2023) identify many challenges related to the attribution of roles and responsibilities, revenue assignment, and fiscal transfers and borrowing.

The perception that institutions lack transparency, fairness, and good governance can undermine people's confidence in government policies and trigger opposition. Sectors that are central to mitigation efforts—such as energy, extractives, land use, and forestry—are also common sites of corruption (OECD 2016; Sovacool 2021; Tacconi and Williams 2020). There is also evidence of corruption in renewable energy technologies, such as hydropower (Pavlakovič et al. 2022; Scudder 2008), wind power (Gennaioli and Tavoni 2016), and solar power (Dvořák et al. 2017). Corruption undermines trust in and support of government and increases inefficiency and costs, which can exacerbate citizens' concerns (Kulin and Johansson Sevä 2021). Research shows that exposure to corruption is a limiting factor for climate policy and is associated with higher emissions (Lamb and Minx 2020). People's support for policies is shaped by their perception of a government's track record, and citizens of countries with poor governance and widespread corruption are less trusting of public policy promises and institutions' capacity to deliver in the public interest. Empirical evidence from Indonesia, for example, shows that opposition to fossil fuel subsidy reform is directly linked to local perceptions of corruption (Kyle 2018).

Administrative capacity, civic consultation and responsiveness, and sector governance norms are key determinants of the effectiveness of climate policy and influence policy instrument choice (Lo 2015, 2021a, 2021b; Macaspac Hernandez 2021). For example, countries with strong institutional capacity tend to favor regulatory over fiscal climate policies (Hughes and Urpelainen 2015). Countries with weak institutions are less likely to be able to introduce complex instruments and policies, such as emission trading systems, technology-focused research and development, or sophisticated targeting systems to protect vulnerable populations. Chapter 3 of this book focuses on how the sequence of climate policies should not only consider capacity constraints but also be designed to build this capacity over time.

Interests diverge

Even when climate policies have positive economic impacts in the aggregate, they may still have large distributional impacts. This is the key focus of chapter 4. In general, opposition can be expected from those who benefit from existing institutions and policies, and who therefore have a vested interest in maintaining the status quo. Similarly, support can be expected from those who stand to benefit from policy change. The reality, however, is often more complex because actors can have multiple, sometimes conflicting, interests. Uneven distributional impacts, such as consumption impacts, sectoral effects on skills and labor, or their spatial dimension, can lead negatively affected actors to prioritize economic over other interests.

Actors involved in or benefiting from fossil fuel extraction and production, energy-intensive industries, or deforestation tend to oppose climate policy. Fossil fuel subsidies and other incentives serve to entrench such interests, sometimes indirectly (box 1.2), especially in the energy sector (Skovgaard and van Asselt 2018). Studies show that when the coal, gas, or oil industry plays a disproportionate role in the power sector, local economy, or exports, opposition from these industries and the spatial effects of climate

BOX 1.2

The challenge of indirect carbon pricing: Hidden incentives and interests

Governments use a variety of instruments to price carbon. The World Bank's annual *State and Trends of Carbon Pricing* reports on progress in this domain (World Bank 2023b). The choice of instrument and policy design depends on the policy objectives and national circumstances, including the political economy. For example, emissions trading systems (ETSs) have often seemed politically easier to introduce than carbon taxes, especially when free permit allocation has protected big emitters against a large increase in costs (World Bank, forthcoming). But direct pricing of emissions through carbon taxes or ETSs does not paint a complete picture of the price incentives facing actors. For example, a fuel excise tax provides a carbon price signal, even though it is not necessarily proportional to a product's relative emissions. Such indirect carbon pricing policies are primarily implemented for purposes other than climate mitigation, such as raising revenue. Fossil fuel subsidies—prevalent across countries—effectively constitute a negative indirect carbon price, counteracting the positive price signal from direct and indirect carbon pricing instruments.

The magnitude of indirect carbon pricing policies dwarf those of direct carbon pricing. In 2022, governments collected almost US$100 billion in revenue from ETSs and carbon taxes, but fossil fuel excise taxes and subsidies were worth over US$1 trillion. An Organisation for Economic Co-operation and Development report estimating net effective carbon rates across 71 countries finds that implicit or indirect carbon prices set by fuel taxes are generally much higher than those set by carbon taxes or ETSs (OECD 2022a).[a] With a weighted average indirect carbon price applied by fossil fuel taxes in 2021 three times the average carbon price set by carbon taxes and ETSs, fuel excise taxes account for about three-quarters of the total positive carbon price. Failure to account for the impact of indirect carbon prices (particularly fossil fuel subsidies) can be misleading and obscure price incentives (Pryor et al. 2023). For example, many countries, including France and Uruguay, have introduced explicit carbon taxes while also reducing other energy taxes to smooth the shock on total energy prices and make the introduction of a new tax more politically acceptable.

Source: J. Pryor, based on World Bank 2023b.

a. The net effective carbon rate includes the carbon price applied by direct carbon pricing instruments (ETSs and carbon taxes) and fuel excise taxes minus fossil fuel subsidies across 71 countries.

change action can create challenges for climate policy making (Fankhauser, Gennaioli, and Collins 2015; Lachapelle and Paterson 2013; Lamb and Minx 2020). Similarly, dominant agriculture and forestry interests can obstruct or undermine climate mitigation efforts (Angelsen et al. 2012; Hochstetler 2020). In addition, policy makers have their own interests that are shaped by things such as their specific responsibilities, their relations to other stakeholders, or lobbying (box 1.3).

Environmental organizations and low-carbon industries, in contrast, have an interest in—and are therefore more likely to support—climate policies. Thus, policy sequencing (explored in detail in chapter 3) can help governments foster supportive interest groups. For example, green innovation and industrial policies can help grow political support coalitions and reduce the cost of low-carbon technologies, increasing the support base for broader climate policies.

Actors' interests are contingent on policy choice and design. For example, labor unions have an interest in protecting jobs. If governments adopt a just transition approach by including active labor market policies and compensation schemes in their climate reforms, labor unions are more likely to support them. Similarly, civic interest groups and the general public may oppose or support climate policy, depending on cost-of-living or distributional outcomes (Dorband et al. 2019; Sovacool 2017). For this reason, the public often favors "pull" (subsidies

BOX 1.3

Policy makers have their own interests

Decision-makers in the political sphere exhibit diverse interests, influenced by factors like their specific responsibilities, relations to other stakeholders, or exposure to lobbying (Jakob et al. 2020). These factors can lead to situations in which the best interests of the public may not align with the individual incentives and rational self-interest that shape the behavior of policy makers (Persson and Tabellini 2002).

Policy makers have interests that affect whether they support or oppose climate policies, notably (re)election to public office. Although it is generally expected that political decision-makers will respond positively to an increased prominence of, and public demand for, climate action (Schaffer, Oehl, and Bernauer 2022), they may avoid or delay action when policies could cost them at the polls or threaten political funding channels (Furceri, Ganslmeir, and Ostry 2021). A group's political power is therefore a decisive aspect in shaping a policy maker's interest, which becomes key in climate politics: the main beneficiaries of strengthened mitigation efforts are not born yet and therefore cannot influence policy makers' interests and current decision-making (Persson and Tabellini 2002). In Europe, green political parties have played an important role in driving the climate agenda and have reaped the benefits at the polls as public demand for climate action increases (McBride 2022). However, examples from Germany, the Netherlands, and Romania, where approval ratings for right-wing populist parties are on the rise, partly in response to government climate action, show that it can work the other way, too.

Lobbying efforts and the power dynamics of interest groups can also affect policy makers' interests and sway policy decisions. The success of these lobbying efforts predominantly depends on the political power of involved groups, determined by aspects such as representation, resources, or networks. For example, highly institutionalized groups like farmers often have greater influence on the political process than does the general public, which is less organized and may therefore be unable to effectively communicate its interests (Persson and Tabellini 2002). As a result, the relationship between policy decisions and public opinion is not always straightforward. For example, Finland collected public opinion on 99 energy and climate policy measures through an online survey to inform its National Energy and Climate Strategy for 2030, yet the resulting policy outcomes did not reflect the survey results (Kinnunen 2021).

The rational self-interest of policy makers can be another driver of decision-making in the political process. In the absence of a coherent climate governance strategy and enabling institutional architecture and capacity, conflicting interests among policy makers can be difficult to overcome, especially when their interests extend beyond their political office. The "revolving doors" phenomenon—whereby individuals move from public office to private companies and vice versa—is often identified as a risk factor. Those advising decision-makers, including scientists and economists, also have interests of their own, such as maintaining influence in policy making, which can make them vulnerable to pressure to legitimize political policy preferences (Geden 2015). For example, some suggest that models that rely heavily on negative emissions technologies despite high costs, low co-benefits, and uncertain feasibility offer a backstop to political interests that want to delay action (Beck and Mahony 2018; Honegger and Reiner 2018; Keyßer and Lenzen 2021; Otto et al. 2021).

and incentives) over "push" (taxes and fines) policies (Drews and van den Bergh 2016)—see figure 1.3. Policy design can be tailored to the interests of different actors to increase support or acceptability (Wicki, Fesenfeld, and Bernauer 2019).

Ideas are a battleground

Actors do not act only in their own political or economic interests; they also pursue and respond to ideas, making the policy process and communication a key part of the challenge. Stakeholder and public knowledge, opinions, and perceptions are key drivers of climate policy decisions, especially in competitive electoral democracies (Drews and van den Bergh 2016). Public knowledge about the effects of a policy is key because refusal can often be

FIGURE 1.3. Share of respondents who supported different climate change policies across 28 countries in 2022

	Asia Pacific	Australia	China	India	Indonesia	Japan	Malaysia	Philippines	Singapore	Korea, Rep.	Thailand	Viet Nam
Carbon pricing		52	55	67	46	39	58	68	58	60	60	74
Subsidies to low-carbon technologies/renewables		65	65	66	62	45	66	73	68	61	69	71
Regulation limiting emissions		52	50	64	44	31	53	60	48	52	56	64

	Americas	Argentina	Brazil	Canada	Colombia	Mexico	United States	Middle East	Egypt, Arab Rep.	Saudi Arabia
Carbon pricing		41	62	50	53	58	44		43	46
Subsidies to low-carbon technologies/renewables		56	65	62	69	69	53		55	57
Regulation limiting emissions		48	50	50	54	56	43		35	41

	Europe	France	Germany	Italy	The Netherlands	Norway	Poland	Spain	Türkiye	United Kingdom
Carbon pricing		44	29	45	40	32	29	44	56	41
Subsidies to low-carbon technologies/renewables		58	54	65	55	50	60	65	61	62
Regulation limiting emissions		44	35	40	38	34	33	43	48	44

Source: World Bank staff calculations, based on data from Dabla-Norris et al. 2023.
Note: Each row in this figure shows the share of favorable responses in each country to the following questions: "Thinking about all of the impacts of a carbon pricing policy, to what extent do you support or oppose such a policy in your country?"; "Thinking about all the impacts of a subsidy to renewable energy, to what extent do you support or oppose this policy in your country?"; and "Thinking about all of the impacts of regulation, to what extent do you support or oppose this policy in your country?," respectively.

traced back to issues of complexity and a lack of understanding rather than to interest-based rejection. For example, studies from Germany show that 62 percent of respondents feel rather or very poorly informed about the carbon pricing scheme and largely overestimate its negative financial effects (Eßler et al. 2023). The interplay between the ideas and interests of the public, decision-makers, influential stakeholders, and political leaders is a standout issue. Political leaders and influential stakeholders can have particular impact, as seen in Brazil, Mexico, the Philippines, and the United States (Lachapelle and Paterson 2013; Marquardt, Oliveira, and Lederer 2022). However, public ideas are also important, especially when they lead to mobilization for or against policies, as demonstrated by the *Gilets Jaunes* protests in France and the global Fridays for Future movement.

Ideologies and worldviews tend to overshadow subjective climate change knowledge, education level, and demographics (Hornsey et al. 2016; McCright, Dunlap, and Marquart-Pyatt 2016). Survey data show that concerns about climate change are high and that the public generally supports climate action, albeit with large variation in support for different climate policies (Ipsos 2022; OECD 2022b) and across nationalities, socioeconomic groups, and educational attainment levels (ISEAS–Yusof Ishak Institute 2022). When comparing public concern about climate change across countries, commitment to democratic values is an important predictor (Lewis, Palm, and Feng 2019). Although ideational factors can coalesce or contribute to making climate change acceptance or denial an intergroup identity issue, polarizing public opinion, their effects differ across contexts. For example, education has positive effects on pro-climate beliefs at low and middle levels of development (Czarnek, Kossowska, and Szwed 2021). At higher levels, however, this effect declines. Social consensus on the reality of climate change and the need for action can also mediate identity- and ideology-based denialism (Goldberg et al. 2020). Trust—in government, science, and peers—is another important mediator that can influence support for climate policies (Drews and van den Bergh 2016; Jagers, Löfgren, and Stripple 2010; Kitt et al. 2021; Kulin and Johansson Sevä 2021; Lamb and Minx 2020).

As with public opinion, ideology and information also shape policy makers' climate beliefs, ideas, and interests (Elgin 2014). Policy making can be especially difficult when ideological polarization exists among decision-makers (Rietig 2019). Policy makers are also often targeted by networks of think tanks and experts funded by—and producing research to the benefit of—vested interests (Franta 2021; Plehwe 2014). This targeting is concerning, because policy makers often rely on scientific or economic ideas to legitimize or change their policy choices (Satoh, Nagel, and Schneider 2022). To support evidence-based policy making and prevent undue influence by vested interests, increased transparency in policy research is vital. Scientific and other research bodies, especially those with a public mandate and transparent funding, can play an important role in informing policy and holding policy makers to account. For example, the United Kingdom's Committee on Climate Change has been instrumental in shaping climate policy (Averchenkova, Fankhauser, and Finnegan 2021). International scientific bodies, agencies, and nongovernmental organizations—such as the Intergovernmental Panel on Climate Change and the International Energy Agency—can also inform policy making by providing research and policy advice.

Although belief in climate change is an important determinant of support for climate action in general, public acceptability of specific policies depends more on design and communication (Hornsey et al. 2016). People often lack the necessary knowledge and firsthand experience to make informed decisions. Moreover, because they have limited time and resources to weigh the costs and benefits of complex policy issues, they rely on trusted actors to make decisions for them (Kitt et al. 2021; Terwel et al. 2010). The public often bases its trust in policy makers and government officials on their perceived competence and integrity, as well as on the extent to which their values align (Kitt et al. 2021). Research in psychology has shown that people are more likely to accept information when it comes from a communicator whom they perceive to be an expert and to have no additional motives for communicating the information (Kelman and Hovland 1953). Policy design, framing, and communication that address public concerns and build trust can help increase public support for climate action. As shown in figure 1.4, perceptions of fairness and effectiveness are the strongest determinants of public support for climate policies (Bergquist et al. 2022). Research shows that focusing policy communication on co-benefits and appealing to values such as community and fairness can build support for climate policy and circumvent ideological obstacles (Bain et al. 2016). Chapter 5 of this book explores how public engagement, policy processes, and communication can facilitate acceptance of and support for climate policies.

Influence has many avenues

Actors' ability and means to influence policy making ultimately determine whose interests and ideas are reflected in policies (box 1.4). Power struggles take place in a range of arenas—from news media to behind-doors lobbying—and at all stages of the policy-making process—from framing issues and solutions, to designing policies and institutions, to interpreting and implementing policies (Morrison et al. 2019). Power dynamics skewed toward incumbents with an interest in maintaining the status quo can derail or delay climate policy. Understanding the balance, source, and mode of influence between actors helps address power imbalances and makes policy processes and outcomes more inclusive and representative (chapter 4 addresses this topic).

FIGURE 1.4. Relationship between determinants and public opinion about climate change taxes and laws

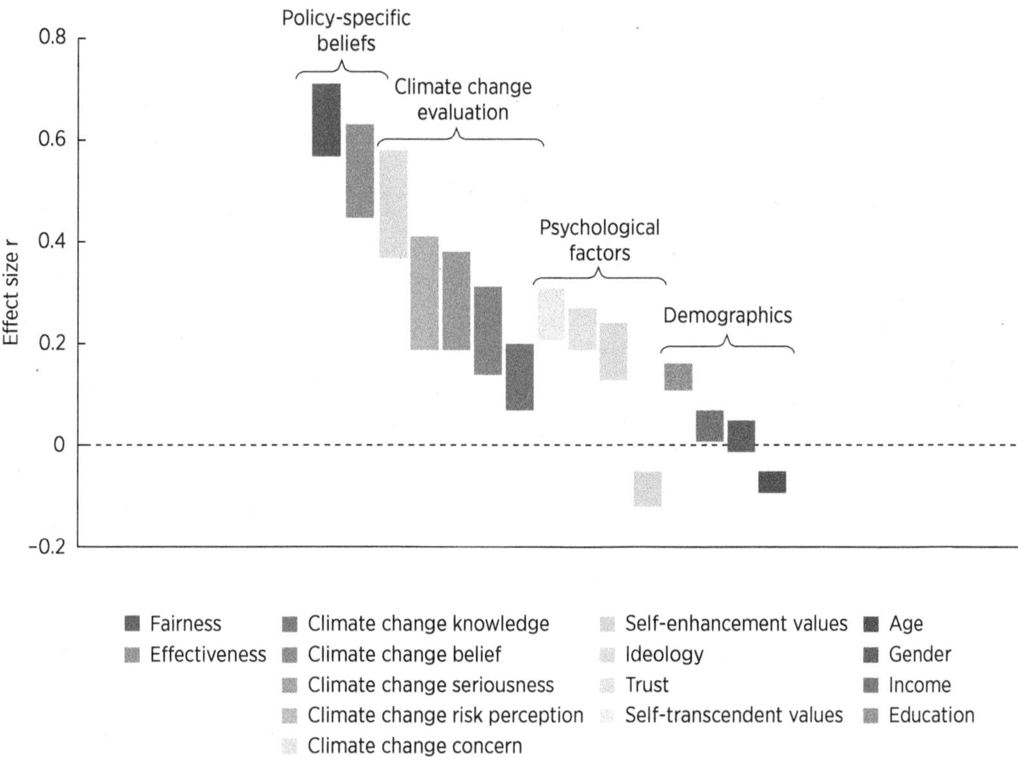

Source: Bergquist et al. 2022.
Note: This figure shows the result of a meta-analysis based on 51 articles incorporating 89 data sets from 33 countries with a total of 119,465 participants. The y-axis measures the importance of a factor (for example, perception of fairness) on a measure of public opinion about climate change taxes and laws (for example, opinion regarding a reform that would increase gasoline prices and reduce public transportation prices).

BOX 1.4

Mapping out actors' interests, ideas, and influence

A common stakeholder analysis approach uses a matrix tool, sometimes called a power-interest matrix, to map out actors' power and preferences. Within the 4i Framework, *power* refers to actors' relative *influence*; their preferences depend on their *interests* and *ideas* (figure B1.4.1).

- *Key actors* (upper right in the figure) have the most influence and have strong preferences related to a specific policy issue or reform. Understanding the interests and ideas underlying their preferences and sources of influence is essential.
- *Latent actors* (upper left) have influence but have no strong preference relating to the policy issue. They might be *context setters*—including institutional bodies, such as regulators—that are not actively involved but whose implicit support is required to pass or implement the policy.
- *Marginalized actors* (lower right) have strong preferences, or may be significantly affected by the policy reform, but have limited influence over other actors or in policy processes. They can quickly move into the upper right quadrant when mobilized—for example, during mass protests or strikes.
- *Apathetic actors* (lower left) have limited influence and weak preferences related to the policy issue or are disengaged.

(Continued)

BOX 1.4
Mapping out actors' interests, ideas, and influence (continued)

FIGURE B1.4.1. **Power-interest matrix**

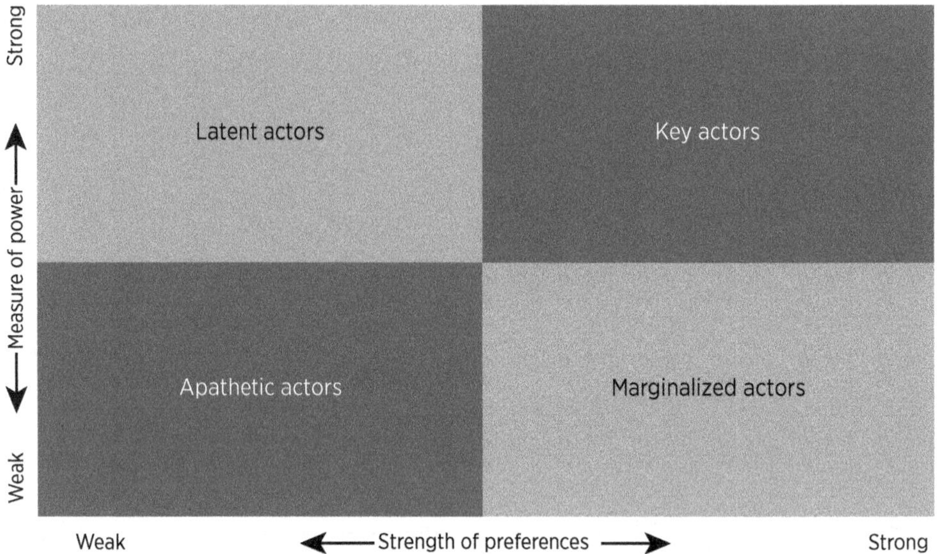

Source: Original figure prepared for this report.

Economic and political incumbents use access, resources, and position to influence policy decisions. Lobbying plays a big role, and incumbent industry has an advantage in terms of organization, resources, and access to decision-makers. In the United States, fossil fuel and transportation companies, utilities, and affiliated trade unions spent at least US$2 billion on climate lobbying between 2010 and 2016 (Brulle 2018; Culhane, Hall, and Roberts 2021; Meng and Rode 2019). In Finland, lobbies have successfully shaped climate policy by directly (and discreetly) lobbying politicians and participating in policy processes (Vesa, Gronow, and Ylä-Anttila 2020).

Incumbent industries can effectively oppose, or sometimes ignore, climate policies. They can oppose policies through noncompliance, particularly when enforcement is costly, information asymmetries exist, and regulation is weak. They can also secure formal or informal exemptions, as attested by the large numbers evading air pollution, gas flaring, and fuel standard regulations in various countries (Cao et al. 2021; Gupta, Saksena, and Baris 2019; Korppoo 2018; Ya'u, Saad, and Mas'ud 2021). Quashing implementation through insufficient budget support is also common, highlighting challenges for designing enforceable policies that are resilient to change or fracture within government. For example, Mexico's General Law on Climate Change, Ley General de Cambio Climático, hailed as one of the world's strongest climate laws when it was passed in 2012, has been weakened by its lack of concrete and timely implementation mechanisms, low support among elected leaders, and lack of budgetary support for climate institutions (Averchenkova 2020; Vance 2012).

Those with less direct influence over policy making or policy makers—notably civil society and civic groups—have used alternative strategies, such as mass mobilization and protests. Outsiders with few resources or little direct access to policy making can use several mechanisms to influence policy. Swaying public knowledge of and demand for climate action is one way to change incentives facing policy makers concerned with electoral support (Newell 2021), and public mobilization has promoted action and urgency (Cheon and Urpelainen 2018; Fisher and Nasrin 2021; Piggot 2018). Indeed, international and local movements, campaigns, and coordination—such as Fridays for Future and Extinction Rebellion—have become increasingly influential (Ginanjar and Mubarrok 2020; Li, Trencher, and Asuka 2022; Marquardt 2020; Rayner 2021).

Litigation is another strategy that has gained in both momentum and impact, especially among people who face barriers to direct participation in or influence over policy processes (Setzer and Vanhala 2019). The number of climate change-related cases targeting governments, private sector actors, and financial institutions has more than doubled globally since 2015 (Setzer and Higham 2022). Some cases—such as that brought by the Dutch Urgenda Foundation—have used a country's constitution or human rights law to challenge the adequacy of existing climate policy to increase government ambition (Wewerinke-Singh and McCoach 2021). Others have used litigation to challenge fossil fuel exploration licenses, environmental impact assessments, and the transparency of decision-making or information, for example, on climate risks; still others have brought corporate liability and fiduciary duty cases against private sector actors (Setzer and Higham 2022). For example, projects for a third runway at London's Heathrow Airport were ruled illegal because the plans did not adequately consider the UK government's climate commitments (Carrington 2020).

Civic actors' influence in climate policy making and implementation depends on, and is often limited by, other political economy factors. In particular, the relative power and influence of vested interests alongside increasing state-sanctioned intimidation and violence limit civic engagement. Environmental defenders tend to face high personal and physical risk, despite relying primarily on nonviolent means of action. In 2020, the global Environmental Justice Atlas identified 2,743 cases of environmental conflict. Among actors taking some form of civic action, including litigation, protest, or mobilization, 18 percent had experienced physical violence; 13 percent had been assassinated; and about 20 percent faced criminalization of dissent through imprisonment, restricted rights, or prosecution without clear charges (Scheidel et al. 2020). Intimidation is not, of course, the only influence; misinformation campaigns and strategic communications by vested interests are especially potent in limiting or redirecting civic engagement (Supran and Oreskes 2017). Around the world, spaces for civic engagement and media freedom, or the protection of these spaces, are being closed, with implications (intentional or otherwise) for environmental civic action and thus on the design of climate policies.

In a nutshell: A dynamic strategy to progressively reduce constraints and build support for climate policies

This book finds that the political economy is not written in stone and should be considered a changing and malleable constraint. When prioritizing and sequencing climate policies, and when designing policy processes and climate policies themselves, governments can apply a dynamic approach to the political economy challenges to climate governance. This dynamic lens leads to the following three key messages.

First, governments should aim to move from opportunistic or unstable to enduring and strategic climate institutions. Country-level political economy dynamics—specifically climate policy narratives and political polarization—determine the best approach to climate institutions in different countries. The climate institutions that are a "good fit" for the political economy today can pave the way for more strategic climate institutions tomorrow. Climate change framework laws, long-term strategies, and just transition frameworks and principles are key strategic climate institutions that can fundamentally alter the political economy of climate policies.

Second, governments should select and sequence policies on the basis of dynamic rather than static efficiency, considering how they feed back on the political economy and balancing short-term feasibility with long-term objectives. The lowest-cost option from a purely economic perspective may well lead to political backlash and create higher costs in the future, whereas choosing a more expensive policy today might be more dynamically efficient if it shifts the political economy to make it easier to implement more efficient policies later. By strategically selecting and sequencing policies, governments can build institutional capacity and create policy beneficiaries who will support further policy action. Governments can also offer firms and people affordable options to substitute away from fossil fuels. They can also leverage reinforcing policy feedback processes and target tipping points in the broader socio-technical-political system to accelerate transformational change. These tipping points, which can be technological, social and behavioral, or political, are key to accelerating decarbonization. Strategically selecting and sequencing feasible policies does not mean climate progress will be slow.

Third, policy process and design need to consider the political economy, including concentrated distributional impacts and the need for policy legitimacy. Climate policies have heterogenous distribution implications across societal groups, income classes, sectors, occupations, or space; and the variance in impacts is larger within income groups than across income groups. Compensation to protect poor and vulnerable populations is possible and affordable; however, the political economy involves more than distributional impacts, and protecting poor households is not enough to ensure acceptability. Opposition to a policy reform is often triggered by concentrated impacts on well-organized or well-connected groups, such as powerful interest groups, organized workers in key sectors, the urban lower-middle class, carbon-intensive regions, or other societal groups, making complementary policies and compensation more challenging to design and implement. Opposition also often originates from a perceived lack of legitimacy of (or agency in) the policy process. Civic engagement and communication can help build legitimacy and develop working compromises and necessary support by mediating distributional conflict, differences in preferences and priorities, and unequal power dynamics.

Note

1. NewClimate Institute, Climate Policy Database, https://climatepolicydatabase.org.

References

Abbott, K. W. 2017. "Orchestrating Experimentation in Non-state Environmental Commitments." *Environmental Politics* 26 (4): 738–63.

Aklin, M., and M. Mildenberger. 2020. "Prisoners of the Wrong Dilemma: Why Distributive Conflict, Not Collective Action, Characterizes the Politics of Climate Change." *Global Environmental Politics* 20 (4): 4–27. https://doi.org/10.1162/glep_a_00578.

Aldy, J. E. 2018. "Policy Surveillance: Its Role in Monitoring, Reporting, Evaluating and Learning." In *Governing Climate Change: Polycentricity in Action*, edited by A. Jordan, D. Huitema, H. van Asselt, and J. Forster. Cambridge: Cambridge University Press.

Angelsen, A., M. Brockhaus, W. D. Sunderlin, and L. V. Verchot, eds. 2012. *Analysing REDD+: Challenges and Choices*. Jawa Barat, Indonesia: Center for International Forestry Research.

Averchenkova, A. 2020. "Mexico's Framework Legislation on Climate Change: Key Features, Achievements and Challenges Ahead." In *National Climate Change Acts: The Emergence, Form and Nature of National Framework Climate Legislation*, edited by T. L. Muinzer, 93–110. Global Energy Law and Policy Series. Oxford: Hart Publishing. https://www.bloomsburycollections.com/monograph-detail?docid=b-9781509941742&pdfid=9781509941742.ch-004.pdf&tocid=b-9781509941742-chapter4.

Averchenkova, A., S. Fankhauser, and J. J. Finnegan. 2021. "The Impact of Strategic Climate Legislation: Evidence from Expert Interviews on the UK Climate Change Act." *Climate Policy* 21(2): 251–63. https://doi.org/10.1080/14693062.2020.1819190.

Bain, P. G., T. L. Milfont, Y. Kashima, M. Bilewica, G. Doron, R. B. Garðarsdóttir, V. V. Gouveia, et al. 2016. "Co-Benefits of Addressing Climate Change Can Motivate Action around the World." *Nature Climate Change* 6 (2): 154–57. https://doi.org/10.1038/nclimate2814.

Bang, G, J. Hovi, and T. Skodvin. 2016. "The Paris Agreement: Short-Term and Long-Term Effectiveness." *Climate Governance and the Paris Agreement* 4 (3). https://doi.org/10.17645/pag.v4i3.640.

Barrett, S., and A. Dannenberg. 2016. "An Experimental Investigation into 'Pledge and Review' in Climate Negotiations." *Climatic Change* 138: 339–51.

Beck, S., and M. Mahony. 2018. "The Politics of Anticipation: The IPCC and the Negative Emissions Technologies Experience." *Global Sustainability* 1. https://doi.org/10.1017/sus.2018.7.

Beeson, M. 2016. "Environmental Authoritarianism and China." In *The Oxford Handbook of Environmental Political Theory*, edited by T. Gabrielson, C. Hall, J. M. Meyer, and D. Schlosberg. Oxford: Oxford University Press. https://doi.org/10.1093/oxfordhb/9780199685271.013.14.

Bergquist, M., A. Nilsson, N. Harring, and S. C. Jagers. 2022. "Meta-Analyses of Fifteen Determinants of Public Opinion about Climate Change Taxes and Laws." *Nature Climate Change* 12(3): 235–40. https://doi.org/10.1038/s41558-022-01297-6.

Bernstein, S., and M. Hoffmann. 2018. "The Politics of Decarbonization and the Catalytic Impact of Subnational Climate Experiments." *Policy Sciences* 51 (2): 189–211.

Betsill, M., N. K. Dubash, M. Paterson, H. van Asselt, A. Vihma, and H. Winkler. 2015. "Building Productive Links between the UNFCCC and the Broader Global Climate Governance Landscape." *Global Environmental Politics* 15 (2): 1–10.

Biermann, F., N. Kanie, and R. E. Kim. 2017. "Global Governance by Goal-Setting: The Novel Approach of the UN Sustainable Development Goals." *Current Opinion in Environmental Sustainability* 26–27: 26–31.

Bromley-Trujillo, R., J. S. Butler, J. Poe, and W. Davis. 2016. "The Spreading of Innovation: State Adoptions of Energy and Climate Policy." *Review of Policy Research* 33 (5): 544–65.

Brulle, R. J. 2018. "The Climate Lobby: A Sectoral Analysis of Lobbying Spending on Climate Change in the USA, 2000 to 2016." *Climatic Change* 149 (3): 289–303.

Cao, X., Q. Deng, X. Li, and Z. Shao. 2021. "Fine Me If You Can: Fixed Asset Intensity and Enforcement of Environmental Regulations in China." *Regulation and Governance* 16 (4). https://doi.org/10.1111/rego.12406.

Cao, X., and H. Ward. 2017. "Transnational Climate Governance Networks and Domestic Regulatory Action." *International Interactions* 43 (1): 76–102.

Carrington, D. 2020. "Heathrow Third Runway Ruled Illegal over Climate Change." *The Guardian*, February 27, 2020. https://www.theguardian.com/environment/2020/feb/27/heathrow-third-runway-ruled-illegal-over-climate-change.

Chayes, A., and A. H. Chayes. 1995. *The New Sovereignty: Compliance with International Regulatory Agreements*. Cambridge, MA: Harvard University Press.

Cheon, A., and J. Urpelainen. 2018. *Activism and the Fossil Fuel Industry*. Routledge and CRC Press. https://www.routledge.com/Activism-and-the-Fossil-Fuel-Industry/Cheon-Urpelainen/p/book/9781783537549.

Culhane, T., G. Hall, and J. T. Roberts. 2021. "Who Delays Climate Action? Interest Groups and Coalitions in State Legislative Struggles in the United States." *Energy Research and Social Science* 79: 102114. https://doi.org/10.1016/j.erss.2021.102114.

Czarnek, G., M. Kossowska, and P. Szwed. 2021. "Right-Wing Ideology Reduces the Effects of Education on Climate Change Beliefs in More Developed Countries." *Nature Climate* 11 (9): 9–13.

Dabla-Norris, E., T. Helbling, S. Khalid, H. Khan, G. Magistretti, A. Sollaci, and K. Srinivasan. 2023. *Public Perceptions of Climate Mitigation Policies: Evidence from Cross-Country Surveys*. Staff Discussion Note SDN2023/002. Washington, DC: International Monetary Fund. https://www.elibrary.imf.org/view/journals/006/2023/002/006.2023.issue-002-en.xml.

Dorband, I. I., M. Jakob, M. Kalkuhl, and J. C. Steckel. 2019. "Poverty and Distributional Effects of Carbon Pricing in Low- and Middle-Income Countries – A Global Comparative Analysis." *World Development* 115: 246–57. https://doi.org/10.1016/j.worlddev.2018.11.015.

Downs, G., D. M. Rocke, and P. N. Barsoom. 1996. "Is the Good News about Compliance Good News about Cooperation?" *International Organization* 50 (2): 379–406.

Drews, S., and J. C. J. M. van den Bergh. 2016. "What Explains Public Support for Climate Policies? A Review of Empirical and Experimental Studies." *Climate Policy* 16 (7): 855–76. https://doi.org/10.1080/14693062.2015.1058240.

Dvořák, P., S. Martinát, D. Van der Horst, B. Frantál, and K. Turečková. 2017. "Renewable Energy Investment and Job Creation: A Cross-Sectoral Assessment for the Czech Republic with Reference to EU Benchmarks." *Renewable and Sustainable Energy Reviews* 69: 360–68. https://doi.org/10.1016/j.rser.2016.11.158.

Elgin, D. J. 2014. "The Effects of Risk, Knowledge and Ideological Beliefs on Climate Policy Preferences: A Study of Colorado Climate and Energy Policy Actors." *Risk, Hazards and Crisis in Public Policy* 5 (1): 1–21. https://doi.org/10.1002/rhc3.12046.

Eskander, S. M. S. U., and S. Fankhauser. 2020. "Reduction in Greenhouse Gas Emissions from National Climate Legislation." *Nature Climate Change* 10 (8): 750–56. https://doi.org/10.1038/s41558-020-0831-z.

Eßler, J., M. Frondel, S. Sommer, and J. Wittmann. 2023. "CO2-Bepreisung in Deutschland: Kenntnisstand privater Haushalte im Jahr 2022." Discussion Paper, RWI – Leibniz Institut für Wirtschaftsforschung, Essen.

Fankhauser, S., C. Gennaioli, and M. Collins. 2015. "The Political Economy of Passing Climate Change Legislation: Evidence from a Survey." *Global Environmental Change* 35: 52–61. https://doi.org/10.1016/j.gloenvcha.2015.08.008.

Farmer, J. D., C. Hepburn, M. C. Ives, T. Hale, T. Wetzer, P. Mealy, R. Rafaty, S. Srivastav, and R. Way. 2019. "Sensitive Intervention Points in the Post-carbon Transition." *Science* 364 (6436): 132–34.

Fay, M., S. Hallegatte, A. Vogt-Schilb, J. Rozenberg, U. Narloch, and T. Kerr. 2015. *Decarbonizing Development: Three Steps to a Zero-Carbon Future*. Washington, DC: World Bank. http://hdl.handle.net/10986/21842.

Finnegan, J. J. 2022. "Institutions, Climate Change and the Foundations of Long-Term Policymaking." *Comparative Political Studies* 55 (7). https://doi.org/10.1177/00104140211047416.

Fisher, D. R., and S. Nasrin. 2021. "Climate Activism and Its Effects." *WIREs Climate Change* 12 (1): e683. https://doi.org/10.1002/wcc.683.

Fransen, T., W. Northrop, K. Mogelgaard, and K. Levin. 2017. "Enhancing NDCs by 2020: Achieving the Goals of the Paris Agreement." Working Paper, World Resources Institute, Washington, DC. https://www.wri.org/research/enhancing-ndcs-2020-achieving-goals-paris-agreement.

Franta, B. 2021. "Weaponizing Economics: Big Oil, Economic Consultants and Climate Policy Delay." *Environmental Politics* 31 (4): 555–75. https://doi.org/10.1080/09644016.2021.1947636.

Fredriksson, P. G., and E. Neumayer. 2013. "Democracy and Climate Change Policies: Is History Important?" *Ecological Economics* 95: 11–19. https://doi.org/10.1016/j.ecolecon.2013.08.002.

Fritz, V., B. Levy, and R. Ort. 2014. *Problem-Driven Political Economy Analysis: The World Bank's Experience*. Directions in Development Series—Public Sector Governance. Washington, DC: World Bank. http://hdl.handle.net/10986/16389.

Furceri, D., M. Ganslmeir, and J. O. Ostry,. 2021. "Are Climate Change Policies Politically Costly?" IMF Working Paper No. 2021/156, International Monetary Fund, Washington, DC. https://www.imf.org/en/Publications/WP/Issues/2021/06/04/Are-Climate-Change-Policies-Politically-Costly-460565.

Geden, O. 2015. "Policy: Climate Advisers Must Maintain Integrity." *Nature* 521 (7550): 27–28. https://doi.org/10.1038/521027a.

Gennaioli, C., and M. Tavoni. 2016. “Clean or Dirty Energy: Evidence of Corruption in the Renewable Energy Sector.” *Public Choice* 166 (3): 261–90. https://doi.org/10.1007/s11127-016-0322-y.

Ginanjar, W. R., and A. Z. Mubarrok. 2020. “Civil Society and Global Governance: The Indirect Participation of Extinction Rebellion in Global Governance on Climate Change.” *Journal of Contemporary Governance and Public Policy* 1 (1): 41–52. https://doi.org/10.46507/jcgpp.v1i1.8.

Godinho, C., S. Hallegatte, and J. Rentschler. Forthcoming. “What Do We Know about the Political Economy of Climate Policies? A Thematic Review of Institutions, Interests, Ideas, and Influence.” Working paper, World Bank, Washington, DC.

Goldberg, M. H., S. van der Linden, A. Leiserowitz, and E. Maibach. 2020. “Perceived Social Consensus Can Reduce Ideological Biases on Climate Change.” *Environment and Behavior* 52 (5): 495–517. https://doi.org/10.1177/0013916519853302.

Gupta, S., S. Saksena, and O. F. Baris. 2019. “Environmental Enforcement and Compliance in Developing Countries: Evidence from India.” *World Development* 117: 313–27. https://doi.org/10.1016/j.worlddev.2019.02.001.

Hale, T. 2020. “Catalytic Cooperation.” *Global Environmental Politics* 20 (4): 73–98.

Hochstetler, K. 2020. “Political Economies of Energy Transition.” In *Political Economies of Energy Transition: Wind and Solar Power in Brazil and South Africa, Business and Public Policy*, edited by K. Hochstetler. Cambridge: Cambridge University Press. https://www.cambridge.org/core/books/political-economies-of-energy-transition/political-economies-of-energy-transition/7B916CDBCF99A5ABF7F41C509F7A82B0.

Honegger, M., and D. Reiner. 2018. “The Political Economy of Negative Emissions Technologies: Consequences for International Policy Design.” *Climate Policy* 18 (3): 306–21. https://doi.org/10.1080/14693062.2017.1413322.

Hornsey, M. J., A. Harris, P. G. Bain, and K. S. Fielding. 2016. “Meta-Analyses of the Determinants and Outcomes of Belief in Climate Change.” *Nature Climate Change* 6 (6): 622–26. https://doi.org/10.1038/nclimate2943.

Hudson, D., and A. Leftwich. 2014. *From Political Economy to Political Analysis.* DLP Research Paper 25. Birmingham: Developmental Leadership Program. https://dlprog.org/publications/research-papers/from-political-economy-to-political-analysis/.

Hughes, L., and J. Urpelainen. 2015. “Interests, Institutions and Climate Policy: Explaining the Choice of Policy Instruments for the Energy Sector.” *Environmental Science and Policy* 54: 52–63. https://doi.org/10.1016/j.envsci.2015.06.014.

IPCC (Intergovernmental Panel on Climate Change). 2022. *Climate Change 2022: Mitigation of Climate Change*. Working Group III contribution to the Sixth Assessment Report of the Intergovernmental Panel on Climate Change. Geneva: IPCC. https://www.ipcc.ch/report/sixth-assessment-report-working-group-3/.

IPCC (Intergovernmental Panel on Climate Change). 2023. *Climate Change 2023. Synthesis Report. Contribution of Working Groups I, II and III to the Sixth Assessment Report of the Intergovernmental Panel on Climate Change*. Geneva: IPCC. https://doi.org/10.59327/IPCC/AR6-9789291691647.

Ipsos. 2022. “New Report Examines People’s Attitudes Towards Climate Change and How They Translate into Action.” News release, June 29, 2022. https://www.ipsos.com/en-uk/new-report-examines-peoples-attitudes-towards-climate-change-and-how-they-translate-action.

ISEAS–Yusof Ishak Institute. 2022. *The Southeast Asia Climate Outlook: 2022 Survey Report.* Singapore: ISEAS–Yusof Ishak Institute. https://www.iseas.edu.sg/wp-content/uploads/2022/08/2022-CCSEAP-Report-7-Sept-final.pdf.

Jackson, G. 2009. *Actors and Institutions.* SSRN Scholarly Paper. Rochester, NY. https://papers.ssrn.com/abstract=1408664.

Jagers, S. C., Å. Löfgren, and J. Stripple. 2010. “Attitudes to Personal Carbon Allowances: Political Trust, Fairness and Ideology.” In *Personal Carbon Trading*, edited Y. Parag and T. Fawcett. Routledge.

Jakob, M., C. Flachsland, J. C. Steckel, and J. Urpelainen. 2020. “Actors, Objectives, Context: A Framework of the Political Economy of Energy and Climate Policy Applied to India, Indonesia and Vietnam.” *Energy Research and Social Science* 70: 101775. https://doi.org/10.1016/j.erss.2020.101775.

Kanie, N., S. Bernstein, F. Biermann, and P. M. Haas. 2017. "Introduction Global Governance through Goal Setting." In *Governing through Goals*, edited by N. Kanie and F. Biermann, 1–28. Cambridge, MA: MIT Press.

Kelley, J. G. 2017. *Scorecard Diplomacy: Grading States to Influence their Reputation and Behavior.* Cambridge: Cambridge University Press.

Kelman, H. C., and C. I. Hovland. 1953. "'Reinstatement' of the Communicator in Delayed Measurement of Opinion Change." *Journal of Abnormal and Social Psychology* 48 (3): 327–35. https://psycnet.apa.org/doi/10.1037/h0061861.

Keohane, R. O., and M. Oppenheimer. 2016. "Paris: Beyond the Climate Dead End through Pledge and Review?" *Politics and Governance* 4 (3): 142–51.

Keyßer, L. T., and M. Lenzen. 2021. "1.5 °C Degrowth Scenarios Suggest the Need for New Mitigation Pathways." *Nature Communications* 12 (1): 2676. https://doi.org/10.1038/s41467-021-22884-9.

Kinnunen, M. 2021. "Weak Congruence Between Public Opinion and Policy Outcome in Energy and Climate Policy – Is There Something Wrong with Finnish Democracy?" *Energy Research and Social Science* 79: 120014.

Kitt, S., J. Axsen, Z. Long, and E. Rhodes. 2021. "The Role of Trust in Citizen Acceptance of Climate Policy: Comparing Perceptions of Government Competence, Integrity and Value Similarity." *Ecological Economics* 183: 106958. https://doi.org/10.1016/j.ecolecon.2021.106958.

Korppoo, A. 2018 "Russian Associated Petroleum Gas Flaring Limits: Interplay of Formal and Informal Institutions." *Energy Policy* 116: 232–41. https://doi.org/10.1016/j.enpol.2018.02.005.

Kulin, J., and I. Johansson Sevä. 2021. "Who Do You Trust? How Trust in Partial and Impartial Government Institutions Influences Climate Policy Attitudes." *Climate Policy* 21 (1): 33–46. https://doi.org/10.1080/14693062.2020.1792822.

Kyle, J. 2018. "Local Corruption and Popular Support for Fuel Subsidy Reform in Indonesia." *Comparative Political Studies* 51 (11). https://doi.org/10.1177/0010414018758755.

Lachapelle, E., and M. Paterson. 2013. "Drivers of National Climate Policy." *Climate Policy* 13 (5): 547–71. https://doi.org/10.1080/14693062.2013.811333.

Lamb, W. F., and J. C. Minx. 2020. "The Political Economy of National Climate Policy: Architectures of Constraint and a Typology of Countries." *Energy Research and Social Science* 64: 101429. https://doi.org/10.1016/j.erss.2020.101429.

Lewis, G. B., R. Palm, and B. Feng. 2019. "Cross-National Variation in Determinants of Climate Change Concern." *Environmental Politics* 28 (5): 793–821. https://doi.org/10.1080/09644016.2018.1512261.

Li, M., G. Trencher, and J. Asuka. 2022. "The Clean Energy Claims of BP, Chevron, ExxonMobil and Shell: A Mismatch between Discourse, Actions and Investments." *PLOS ONE* 17 (2): e0263596. https://doi.org/10.1371/journal.pone.0263596.

Lo, K. 2015. "How Authoritarian Is the Environmental Governance of China?" *Environmental Science and Policy* 54: 152–59. https://doi.org/10.1016/j.envsci.2015.06.001.

Lo, K. 2021a. "Authoritarian Environmentalism, Just Transition and the Tension between Environmental Protection and Social Justice in China's Forestry Reform." *Forest Policy and Economics* 131: 102574. https://doi.org/10.1016/j.forpol.2021.102574.

Lo, K. 2021b. "Can Authoritarian Regimes Achieve Just Energy Transition? Evidence from China's Solar Photovoltaic Poverty Alleviation Initiative." *Energy Research and Social Science* 82: 102315. https://doi.org/10.1016/j.erss.2021.102315.

Macaspac Hernandez, A. 2021. *Taming the Big Green Elephant*. Wiesbaden: Springer VS. https://doi.org/10.1007/978-3-658-31821-5_9.

Macy, M. W. 1991. "Chains of Cooperation: Threshold Effects in Collective Action." *American Sociological Review* 56 (6): 730–47.

Marquardt, J. 2020. "Fridays for Future's Disruptive Potential: An Inconvenient Youth Between Moderate and Radical Ideas." *Frontiers in Communication* 5. https://www.frontiersin.org/article/10.3389/fcomm.2020.00048.

Marquardt, J., M. C. Oliveira, and M. Lederer. 2022. "Same, Same but Different? How Democratically Elected Right-Wing Populists Shape Climate Change Policymaking." *Environmental Politics* 31 (5): 777–800. https://doi.org/10.1080/09644016.2022.2053423.

Martinez-Vazquez, J., and F. Zahir. 2023. "Climate Change Strategy and India's Federalism." Working Paper 23-19, International Center for Public Policy, Atlanta.

McBride, J. 2022. "How Green-Party Success Is Reshaping Global Politics." Council on Foreign Relations Backgrounder, May 5, 2022. https://www.cfr.org/backgrounder/how-green-party-success-reshaping-global-politics.

McCright, A. M., R. E. Dunlap, and S. T. Marquart-Pyatt. 2016. "Political Ideology and Views about Climate Change in the European Union." *Environmental Politics* 25 (2): 338–58. https://doi.org/10.1080/09644016.2015.1090371.

Meng, K. C., and A. Rode. 2019. "The Social Cost of Lobbying Over Climate Policy." *Nature Climate Change* 9 (6): 472–76. https://doi.org/10.1038/s41558-019-0489-6.

Michaelowa, A. 2021. "Solar Radiation Modification: A 'Silver Bullet' Climate Policy for Populist and Authoritarian Regimes?" *Global Policy* 12 (S1): 119–28. https://doi.org/10.1111/1758-5899.12872.

Morrison, T. H., W. N. Adger, K. Brown, M. C. Lemos, D. Huitema, J. Phelps, L. Evans, P. Cohen, A. M. Song, R. Turner, T. Quinn, and T. P. Hughes. 2019. "The Black Box of Power in Polycentric Environmental Governance." *Global Environmental Change* 57: 101934. https://doi.org/10.1016/j.gloenvcha.2019.101934.

NewClimate Institute and Climate Analytics. 2019. "Climate Action Tracker: Warming Projections Global Update, December 2019." NewClimate Institute and Climate Analytics. https://climateactiontracker.org/documents/698/CAT_2019-12-10_BriefingCOP25_WarmingProjections GlobalUpdate_Dec2019.pdf.

NewClimate Institute. 2022. *Climate Action Tracker: Warming Projections Global Update (November)*. Cologne: NewClimate Institute. https://climateactiontracker.org/documents/1094/CAT_2022-11-10_GlobalUpdate_COP27.pdf.

Newell, P. 2021. *Power Shift: The Global Political Economy of Energy Transitions*. Cambridge: Cambridge University Press. https://doi.org/10.1017/9781108966184.

Nordhaus, W. 2021. "Dynamic Climate Clubs: On the Effectiveness of Incentives in Global Climate Agreements." *Proceedings of the National Academy of Sciences*, 118 (45): e2109988118.

OECD (Organisation for Economic Co-operation and Development). 2016. *Corruption in the Extractive Value Chain: Typology of Risks, Mitigation Measures and Incentives* Paris: OECD. https://www.oecd-ilibrary.org/development/corruption-in-the-extractive-value-chain_9789264256569-en.

OECD (Organisation for Economic Co-operation and Development). 2022a. *Pricing Greenhouse Gas Emissions: Turning Climate Targets into Climate Action, OECD Series on Carbon Pricing and Energy Taxation*. Paris: OECD Publishing. https://doi.org/10.1787/e9778969-en.

OECD (Organisation for Economic Co-operation and Development). 2022b. *Fighting Climate Change: International Attitudes Toward Climate Policies*. Parus: OECD Publishing https://www.oecd-ilibrary.org/economics/fighting-climate-change-international-attitudes-toward-climate-policies_3406f29a-en.

Otto, D., T. Thoni, F. Wittstick, and S. Beck. 2021. "Exploring Narratives on Negative Emissions Technologies in the Post-Paris Era." *Frontiers in Climate* 3. https://www.frontiersin.org/article/10.3389/fclim.2021.684135.

Pavlakovič, B., A. Okanovic, B. Vasić, J. Jesic, and P. Šprajc. 2022. "Small Hydropower Plants in Western Balkan Countries: Status, Controversies and a Proposed Model for Decision Making." *Energy, Sustainability and Society* 12 (1): 9. https://doi.org/10.1186/s13705-022-00335-7.

Persson, T., and G. Tabellini. 2002. "Political Economics and Public Finance." In *Handbook of Public Economics*," Volume 3, edited by A. J. Auerbach and M. Feldstein, 1549–659. Elsevier. https://doi.org/10.1016/S1573-4420(02)80028-3.

Piggot, G. 2018. "The Influence of Social Movements on Policies That Constrain Fossil Fuel Supply." *Climate Policy* 18 (7): 942–54. https://doi.org/10.1080/14693062.2017.1394255.

Plehwe, D. 2014. "Think Tank Networks and the Knowledge–Interest Nexus: The Case of Climate Change." *Critical Policy Studies* 8 (1): 101–15. https://doi.org/10.1080/19460171.2014.883859.

Pryor, J., P. Agnolucci, C. Fischer, D. Heine, and M. Montes de Oca Leon. 2023. "Carbon Pricing Around the World." In *Data for a Greener World: A Guide for Practitioners and Policymakers*, edited by S. Arslanalp, K. Kostial, and G. Quirós-Romero. Washington, DC: International Monetary Fund. https://www.elibrary.imf.org/display/book/9798400217296/9798400217296.xml.

Rayner, T. 2021. “Keeping It in the Ground? Assessing Global Governance for Fossil-Fuel Supply Reduction.” *Earth System Governance* 8: 100061. https://doi.org/10.1016/j.esg.2020.100061.

Rietig, K. 2019. “The Importance of Compatible Beliefs for Effective Climate Policy Integration.” *Environmental Politics* 28 (2): 228–47. https://doi.org/10.1080/09644016.2019.1549781.

Sabel, C. F., and D. G. Victor. 2017. “Governing Global Problems Under Uncertainty: Making Bottom-Up Climate Policy Work.” *Climatic Change* 144 (1): 15–27.

Sachs, L. E. 2019. “The Paris Agreement in the 2020s: Breakdown or Breakup?” *Ecology Law Quarterly* 46 (1).

Satoh, K., M. Nagel, and V. Schneider. 2022. “Organizational Roles and Network Effects on Ideational Influence in Science-Policy Interface: Climate Policy Networks in Germany and Japan.” *Social Networks* 75: 88–106. https://doi.org/10.1016/j.socnet.2022.01.014.

Schaffer, L. M., B. Oehl, and T. Bernauer. 2022. “Are Policymakers Responsive to Public Demand in Climate Politics?” *Journal of Public Policy* 42 (1): 136–64. https://doi.org/10.1017/S0143814X21000088.

Scheidel, A., D. Del Bene, J. Liu, G. Navas, S. Mingorría, F. Demaria, S. Avila, B. Roy, I. Ertör, L. Temper, and J. Martínez-Alier. 2020. “Environmental Conflicts and Defenders: A Global Overview.” *Global Environmental Change* 63: 102104. https://doi.org/10.1016/j.gloenvcha.2020.102104.

Scudder, T. 2008. “Hydropower Corruption and the Politics of Resettlement.” *Global Corruption Report* 7(4): 96.

Setzer, J., and C. Higham. 2022. *Global Trends in Climate Litigation: 2022 Snapshot*. London: Grantham Research Institute on Climate Change and the Environment. https://eprints.lse.ac.uk/117652/.

Setzer, J., and L. C. Vanhala. 2019. “Climate Change Litigation: A Review of Research on Courts and Litigants in Climate Governance.” *WIREs Climate Change* 10 (3): e580. https://doi.org/10.1002/wcc.580.

Shobe, W. 2020. “Emerging Issues in Decentralized Resource Governance: Environmental Federalism, Spillovers and Linked Socio-Ecological Systems.” *Annual Review of Resource Economics* 12 (1): 259–79. https://doi.org/10.1146/annurev-resource-110319-114535.

Skovgaard, J., and H. van Asselt. 2018. “The Politics of Fossil Fuel Subsidies and Their Reform: An Introduction.” In *The Politics of Fossil Fuel Subsidies and Their Reform*, edited by H. van Asselt and J. Skovgaard, 3–20. Cambridge: Cambridge University Press. https://doi.org/10.1017/9781108241946.003.

Sovacool, B. K. 2017. “Reviewing, Reforming and Rethinking Global Energy Subsidies: Towards a Political Economy Research Agenda.” *Ecological Economics* 135: 150–63. https://doi.org/10.1016/j.ecolecon.2016.12.009.

Sovacool, B. K. 2021. “Clean, Low-Carbon but Corrupt? Examining Corruption Risks and Solutions for the Renewable Energy Sector in Mexico, Malaysia, Kenya and South Africa.” *Energy Strategy Reviews* 38: 100723. https://doi.org/10.1016/j.esr.2021.100723.

Steurer, R., and C. Clar. 2015. “Is Decentralisation Always Good for Climate Change Mitigation? How Federalism has Complicated the Greening of Building Policies in Austria.” *Policy Sciences* 48 (1): 85–107. https://doi.org/10.1007/s11077-014-9206-5.

Stokes, L. C. 2020. *Short Circuiting Policy: Interest Groups and the Battle Over Clean Energy and Climate Policy in the American States*. Oxford: Oxford University Press.

Supran, G., and N. Oreskes. 2017. “Assessing ExxonMobil’s Climate Change Communications (1977–2014).” *Environmental Research Letters* 12 (8): 084019. https://doi.org/10.1088/1748-9326/aa815f.

Tacconi, L., and D. A. Williams. 2020. “Corruption and Anti-Corruption in Environmental and Resource Management.” *Annual Review of Environment and Resources* 45 (1): 305–29. https://doi.org/10.1146/annurev-environ-012320-083949.

Terwel, B. W., F. Harinck, N. Ellemers, and D. D. L. Daamen. 2010. “Voice in Political Decision-Making: The Effect of Group Voice on Perceived Trustworthiness of Decision Makers and Subsequent Acceptance of Decisions.” *Journal of Experimental Psychology: Applied* 16 (2): 173–186. https://doi.org/10.1037/a0019977.

UNFCCC (United Nations Framework Convention on Climate Change). 2022. *UNFCCC Standing Committee on Finance. Report on Progress towards Achieving the Goal of Mobilizing Jointly USD 100 Billion per Year to Address the Needs of Developing Countries in the Context of Meaningful Mitigation Actions and Transparency on Implementation.* Bonn: UNFCCC. https://unfccc.int/process-and-meetings/bodies/constituted-bodies/standing-committee-on-finance-scf/progress-report.

UNFCCC (United Nations Framework Convention on Climate Change). 2023. "Technical Dialogue of the First Global Stocktake. Synthesis Report by the Co-facilitators on the Technical Dialogue." Report prepared for the UN Climate Change Conference, United Arab Emirates, November/December 2023. https://unfccc.int/documents/631600.

Urpelainen, J. 2009. "Explaining the Schwarzenegger Phenomenon: Local Frontrunners in Climate Policy." *Global Environmental Politics* 9 (3): 82–105.

van Asselt, H. 2016. "The Role of Non-state Actors in Reviewing Ambition, Implementation, and Compliance under the Paris Agreement." *Climate Law* 6 (1): 91–108.

Vance, E. 2012. "Mexico Passes Climate-Change Law." *Nature*, April 20, 2012. https://doi.org/10.1038/nature.2012.10496.

Vesa, J., A. Gronow, and T. Ylä-Anttila. 2020. "The Quiet Opposition: How the Pro-Economy Lobby Influences Climate Policy." *Global Environmental Change* 63: 102117. https://doi.org/10.1016/j.gloenvcha.2020.102117.

Victor, D. G., K. Raustiala, and E. B. Skolnikoff. 1998. *The Implementation of International Environmental Commitments: Theory and Practice.* Cambridge, MA: MIT Press.

von Stein, J. 2022. "Democracy, Autocracy and Everything in Between: How Domestic Institutions Affect Environmental Protection." *British Journal of Political Science* 52 (1): 339–57. https://doi.org/10.1017/S000712342000054X.

Wewerinke-Singh, M., and A. McCoach. 2021. "The State of the Netherlands v Urgenda Foundation: Distilling Best Practice and Lessons Learnt for Future Rights-Based Climate Litigation." *Review of European, Comparative and International Environmental Law* 30 (2): 275–83. https://doi.org/10.1111/reel.12388.

Wicki, M., L. Fesenfeld, and T. Bernauer. 2019. "In Search of Politically Feasible Policy-Packages for Sustainable Passenger Transport: Insights from Choice Experiments in China, Germany and the USA." *Environmental Research Letters* 14 (8): 084048. https://doi.org/10.1088/1748-9326/ab30a2.

World Bank. 2013. *World Development Report 2014: Risk and Opportunity—Managing Risk for Development.* Washington, DC: World Bank. http://hdl.handle.net/10986/16092.

World Bank. 2017. *World Development Report 2017: Governance and the Law.* Washington, DC: World Bank. https://www.worldbank.org/en/publication/wdr2017.

World Bank. 2023a. *Reality Check: Lessons from 25 Policies Advancing a Low-Carbon Future.* Washington, DC: World Bank. http://hdl.handle.net/10986/40262.

World Bank. 2023b. *State and Trends of Carbon Pricing 2023.* Washington, DC: World Bank. http://hdl.handle.net/10986/13334.

World Bank. 2023c. "Climate Policies with Real-World Results." World Bank Feature Story, September 19, 2023. https://www.worldbank.org/en/news/feature/2023/09/19/climate-policies-with-real-world-results.

World Bank. Forthcoming. *The Political Economy of Carbon Pricing: A Practical Review.*

World Bank Group. 2022. *Climate and Development: An Agenda for Action—Emerging Insights from World Bank Group 2021–22 Country Climate and Development Reports.* Washington, DC: World Bank. http://hdl.handle.net/10986/38220.

Ya'u, A., N. Saad, and A. Mas'ud. 2021. "Validating the Effects of the Environmental Regulation Compliance Scale: Evidence from the Nigerian Oil and Gas Industry." *Environmental Science and Pollution Research* 28 (11): 13570–80. https://doi.org/10.1007/s11356-020-11608-z.

2 Climate Governance

Strategically (Re)Building the Institutional Context for Transition

KEY INSIGHTS

National climate governance is about how states build, use, and adapt formal institutions (organizations and rules) to address climate change by making credible commitments and institutional arrangements that facilitate coordination and cooperation to achieve them.

Governments should aim to move from opportunistic or unstable to enduring and strategic climate institutions. Climate institutions that are a good fit for the political economy pave the way for more strategic climate institutions.

Climate change framework laws, long-term strategies, and just transition frameworks and principles are key strategic climate institutions that can fundamentally alter the political economy of climate policies. Creating a whole-of-economy institutional structure and a shared vision for action can help align institutions, interests, ideas, and influence with the transition.

The previous chapter shows how most countries face political economy constraints, making it difficult to adopt and implement sufficiently stringent climate policies. This chapter focuses on the role of climate governance and institutions, looking at how the political economy shapes the emergence of climate institutions, from often opportunistic or ad hoc structures to more strategic approaches.

Using formal institutions to address climate change on a national level

Governance can be described as the way rules, norms, and actions are structured, sustained, regulated, and held accountable. It is the process through which state and nonstate actors interact to design and implement policies within a given set of formal and informal rules that shape and are shaped by power (World Bank 2017). This process is frequently affected by common problems such as capture, unbalanced resource allocations, and limited stakeholder engagement in decision-making (figure 2.1).

Public institutions are key in shaping the governance process and executing process outcomes. Public institutions include government and public service bodies that structure public affairs, deliver services, and implement regulations, and the rules that govern those bodies. To enable climate policies, institutions need to perform three key functions: enabling credible commitment, inducing coordination, and enhancing cooperation (World Bank 2017)—see table 2.1.

Climate institutions can help establish formal rules and organizations that facilitate the design, implementation, and enforcement of effective climate policies. They include laws, strategies, frameworks, and institutional bodies and organizations that alter the way climate policies are made and enacted (box 2.1). For example, Chile enacted a Climate Change Framework Law in 2022 imposing legally binding climate neutrality by 2050 and establishing principles such as scientific validity, cost-effectiveness, citizen participation, and equity and climate justice. The country also created new climate governance bodies, such as the Council of Ministers for Sustainability and Climate Change and the Scientific Advisory Committee (Grantham Research Institute 2022). These new formal rules and organizational bodies have the potential to fundamentally alter the way the country makes its climate policies.

Governments will need additional capacities and resources to fulfill climate governance functions across institutions. This includes expanding capacities for planning and

FIGURE 2.1. **Governance dimensions and frequent governance problems**

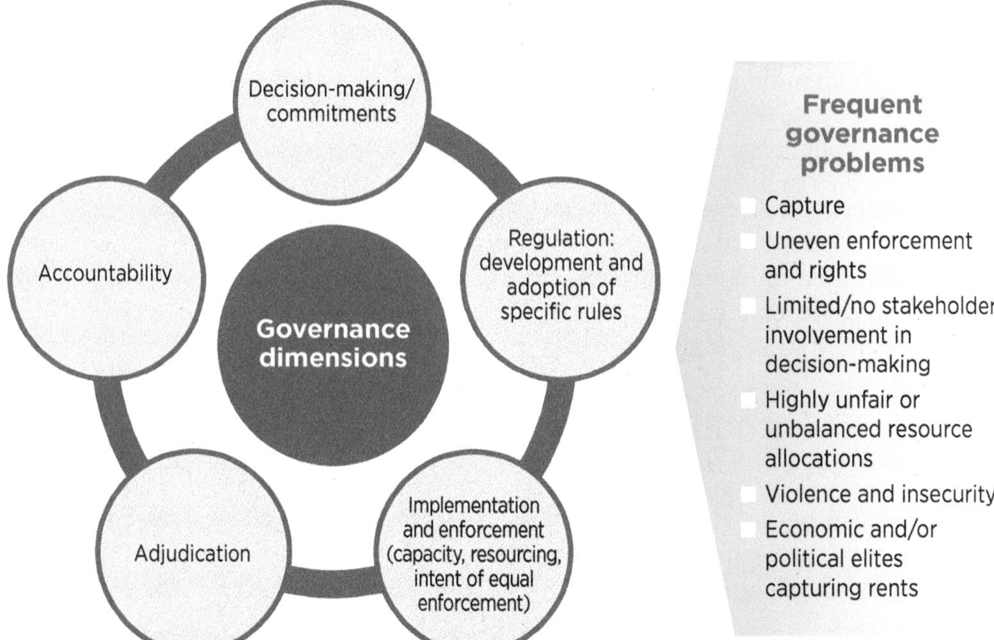

Source: Original figure developed for this report.

TABLE 2.1. Climate governance: Three core functions of institutions

Governance function	Tasks	Examples
Enabling credible commitment	Make credible climate commitments by setting long-term and intermediate targets. Develop clear policy plans that capture scaling, acceleration, adjustment, and enforcement mechanisms. Establish and capacitate governance and financing mechanisms. Establish and capacitate accountability mechanisms.	Regular NDC updates Framework climate legislation and long-term pledges or targets Integrating climate into national development plans LTSs and sectoral targets and plans Regular reviews of policies and regulations Climate institutions and bodies responsible for governance, implementation, and monitoring and evaluation
Inducing coordination	Coordinate actions between actors, across sectors, and between levels by developing and updating systematic strategies and road maps. Collect and make information available to stakeholders. Institute appropriate coordination mechanisms.	Cross-sectoral coordination mechanisms, such as interministerial, interregional, or parliamentary forums LTSs and sectoral transition strategies Emissions reporting Green budget tagging Private-public forums
Enhancing cooperation	Develop and adopt binding rules and regulations. Introduce incentives and disincentives to prevent free riding and generate compliance. Implement engagement and communication strategies that inform, empower, and help secure buy-in from stakeholders, particularly the public. Develop and ensure the necessary capacity, resourcing for implementation, and equal enforcement. Adjudicate when necessary.	Incentive structures with a mix of regulations and pricing instruments for emissions, pollution, pricing, and so on Fiscal policies, including taxes, subsidies, and public investment Communication instruments, such as labeling, public engagement, education and training, and public reporting

Source: Original table developed for this report, based on World Bank 2017.
Note: LTS = long-term strategy; NDC = nationally determined contribution.

managing energy transitions or conducting climate-informed public investment appraisals and developing new functions, such as techniques for monitoring greenhouse gas emissions. Meeting this need will require additional staff and financial resources to deliver new mandates, from intensifying forest protection efforts to enhancing agricultural extension services and assessing priorities in public investment management. Although some governments will face challenges in financing these needs on top of other development imperatives, they will find it easier with social and political consensus that addressing climate change adaptation, mitigation, and just transition expenditures are both a priority and in the interest of economic development.

Lessons from real-world climate institutions

Real-world examples of climate institution-building efforts demonstrate how the political economy can shape the form and functionality of these institutions. Dubash et al. (2021) explore how political economy dynamics—specifically climate policy narratives and

BOX 2.1

The World Bank's Climate Change Institutional Assessment

The Climate Change Institutional Assessment (CCIA) is a framework that helps countries assess the strengths and weaknesses of their institutions to address climate change. It has been applied in more than 50 countries in all regions, at all stages of development, and facing the full range of climate impacts. The CCIA covers five pillars—organization, planning, public finances, subnational governments and state-owned enterprises, and accountability—and 23 topics that address the political economy challenges of climate change (figure B2.1.1). The CCIA focuses on crosscutting, center-of-government institutions, such as planning, finance, economy, and climate-specific agencies, and on mechanisms for engaging with nongovernment stakeholders. An umbrella diagnostic, it has complementary tools—such as PEFA Climate,[a] which identifies the strengths and weaknesses of climate-responsive public financial management, and C-PIMA,[b] which assesses countries' capacity to manage climate-related infrastructure—for in-depth analysis of specific topics. It complements diagnostics that focus on the private, financial, and specific sectors (energy, transportation, agriculture, and so on). Each CCIA produces a set of recommendations that have informed the preparation of nationally determined contributions, long-term solutions, technical assistance, and investments.

FIGURE B2.1.1. **The five pillars of the CCIA**

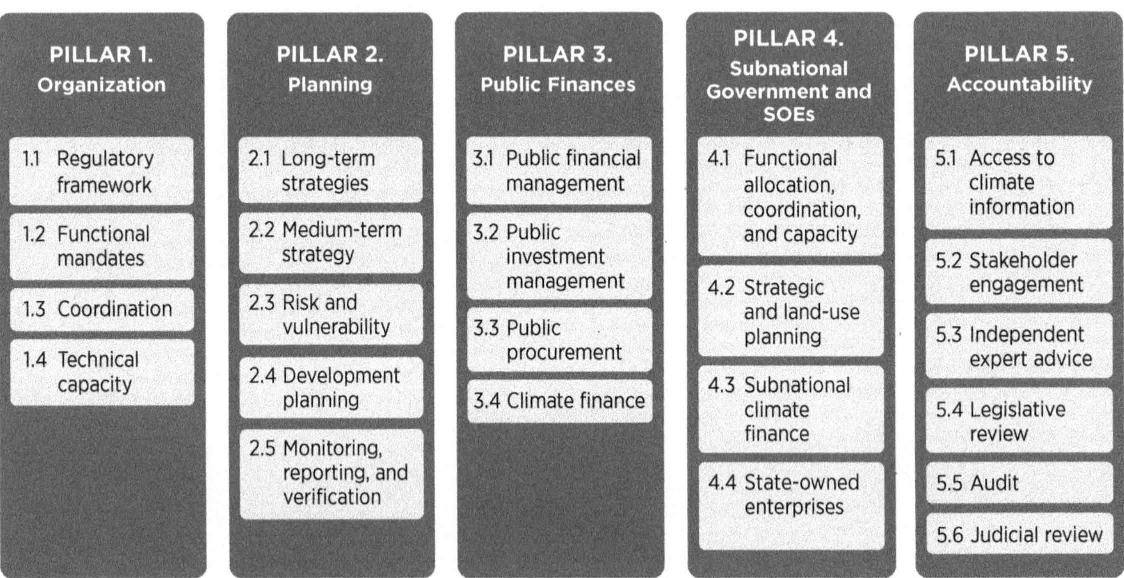

Source: Original figure developed for this report, based on World Bank 2021.
Note: CCIA = Climate Change Institutional Assessment.

a. For more about PEFA Climate (Public Expenditure and Financial Accountability Climate Responsive Public Financial Management Framework), see https://www.pefa.org/resources/climate-responsive-public-financial-management-framework-pefa-climate-piloting-phase.

b. For more about C-PIMA (Climate-Public Investment Management Assessment), see https://infrastructuregovern.imf.org/content/PIMA/Home/PimaTool/C-PIMA.html.

political polarization—shape the form and functioning of climate institutions in different countries. Political narratives reveal dominant policy ideas, which are divided into opposing narratives:

- *Mitigation-centric narratives* emerge where climate change mitigation is already a well-established and high-priority public goal that allows for explicit emissions reduction framing.
- *Embedded climate narratives* are likely when climate is lower on the agenda, so it is easier to use framing that subsumes climate goals under other objectives, such as green growth, energy security, or job creation.

Moreover, the prevailing levels of political polarization are related to the pro- or anti-climate interests of influential actors:

- *High levels of polarization* are likely when there are influential pro- and anticlimate actors, especially if there are distributional conflicts between winners and losers (as is often the case with oil and gas exporters).
- *Low levels of polarization* are likely when most influential actors are pro-climate policy and/or in the case of limited trade-offs or costs for powerful actors or societal goals.

Based on these dimensions, Dubash et al. (2021) identify four types of climate politics (table 2.2) that tend to produce four distinct types of climate governance institutions: opportunistic, strategic, unstable sectoral, and unstable climate institutions.

Strategic climate institutions are more acceptable, durable, and effective when a working political climate consensus exists—that is, when the need for ambitious climate policies is widely accepted in the public discourse—and levels of political contestation are lower. Governments can use strategic climate institutions to fundamentally (re)structure the state response to climate change. These institutions provide an overarching institutional framework that guides climate vision, target setting, decision-making, institutional development, and implementation, bringing together the core governance functions. They can take the form of a climate law or strategy that fulfills the functions of narrative and high-level direction as well as setting out principles for action, mechanisms for upgrading existing institutions to enable and coordinate policy making and implementation, and plans for mobilizing and channeling finance (Sridhar et al. 2022). To be credible, these institutions should include oversight, accountability, and enforcement measures as well as processes that support stakeholder alignment (box 2.2).

TABLE 2.2. Types of climate governance

Interests		**Ideas**	
		Dominant narrative on climate policies	
		Embedded	**Mitigation-centric**
Extent of political polarization of climate policy	Low	Under-the-radar climate politics Opportunistic climate institutions	Climate consensus politics Strategic climate institutions
	High	"Contested sector" politics Unstable sectoral institutions	In-the-crossfire politics Unstable climate institutions

Source: Dubash 2021.

BOX 2.2

How consensus enabled the United Kingdom's strategic climate institutions and 2008 Climate Change Act

The United Kingdom provides an example of climate consensus politics that have enabled and then been advanced through strategic climate institutions, specifically the 2008 Climate Change Act (Averchenkova, Fankhauser, and Finnegan 2021). The act followed almost two decades of climate institution building, from the 1994 UK Programme on Climate Change to the 2000 Climate Change Programme, which established mitigation-centric discourse, bolstered by the country's participation in the 1997 Kyoto Protocol through the European Union and adopting targets under EU climate packages.

The Climate Change Act built on and enhanced mitigation targets and established additional climate institutions, most prominently the independent Climate Change Committee. The absence of a powerful coal lobby—the result of dismantling the coal industry in the 1980s—certainly helped. Other political economy-enabling factors include strong democratic institutions and administrative capacity. The act also favored market-mimicking instruments that align with prevailing market ideology-based preferences, and the Climate Change Committee has played a key role in shaping ideas and policy discourse. The United Kingdom has been able to meet its five-year mitigation targets, with especially deep declines in the power sector.

Sources: Lockwood 2021; World Bank 2023.

By contrast, high political polarization or a contested dominant narrative on climate makes it much more difficult to develop strategic institutions and stable sectoral institutions. When there is greater resistance and reactivity in the political economy, it can be difficult to consolidate governance gains and build climate policy progress; climate institutions may be blocked or rolled back as vested interests work against them. This happened with Brazil's Action Plan for the Prevention and Control of Deforestation in the Legal Amazon. In such a context, even efforts to layer climate institutions in key sectors are at high risk of backlash because the political economy is highly sensitive and reactive to change (box 2.3).

Opportunistic climate institutions tend to emerge when climate politics are under the radar because of climate narratives embedded in other objectives and low levels of political contestation around climate change, as is the case in many low- and middle-income countries. Because climate change is not central to politics, or may be actively kept out of politics, climate change mitigation is tailored to other domestic agendas and layered onto existing institutions—for example, by adding climate change mandates or reporting to existing portfolios, such as energy or transportation (box 2.4). Opportunistic institutions can avoid increasing contestation or polarization, especially in contexts where mitigation could encounter resistance if it is perceived to be at odds with other priorities or discourses. Such institutions often depend on individual champions but can wither when their support wanes, because they are less likely to be institutionalized through legislation. By aligning climate change objectives with national development priorities, opportunistic climate institutions can be an important first step toward developing more established and dedicated climate institutions in the longer term. Lacking frameworks that can align actions across sectors, they also face greater risk of miscoordination, which can result in duplication of efforts or interventions that work at cross-purposes.

BOX 2.3

Political economy barriers to Brazil's Action Plan for the Prevention and Control of Deforestation in the Legal Amazon

The 2004 Action Plan for the Prevention and Control of Deforestation in the Legal Amazon, overseen and implemented by several government bodies, contributed to a 76 percent reduction in the annual deforestation rate between 2005 and 2012. This achievement demonstrates existent and available governance capacity. But the success of these efforts drew backlash from rural and agricultural interests, which exerted considerable pressure on the government and politicians in a macroeconomic context that magnifies revenues from extensive agriculture (Hanusch 2023). Under pressure from these vested interests, budget allocations declined from 2011, regulations weakened—for example, the 2012 revised Forest Code vastly reduced the area required for legal reserves on rural private properties and deaccelerated the implementation of the Environmental Cadaster—and high-level political support dissolved, eroding the capacity of the climate institutions underpinning the anti-deforestation efforts (Hochstetler 2021). By 2020, Brazilian Amazon deforestation rates were at their highest in a decade (figure B2.3.1).

In its first six months, the new administration, which took office early in 2023, demonstrated the importance of political commitment for reinstating public policies in support of the environmental protection agenda. Effective policy changes to date—including strengthening the environmental protection agency Ibama as the authority in charge of combating illegal deforestation and the indigenous agency FUNAI, leading to the demarcation of new indigenous territories, the reactivation of the Amazon fund, and the swift approval of the cross-ministerial action plans to combat deforestation—have led to a significant reduction in deforestation within a few months. These ups and downs of climate policy implementation show that the political economy is at least as important as capacity when it comes to ensuring durable climate action. A consensus of all major stakeholders will now be needed to maintain those initial results and avoid a new political backlash.

FIGURE B2.3.1. **Deforestation in Brazil, 1996–2020**

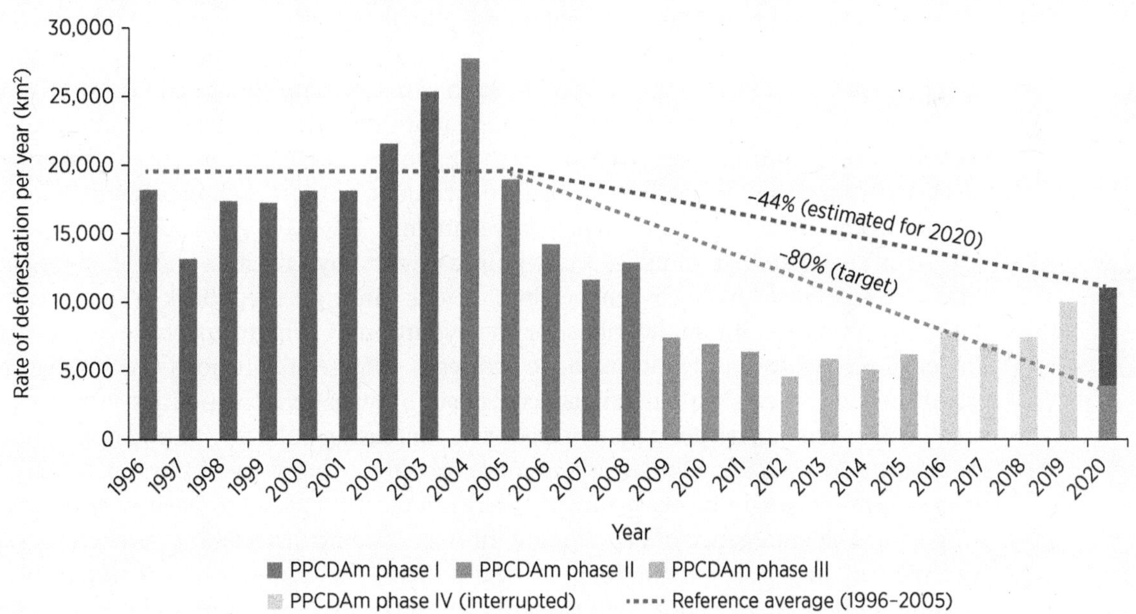

Source: Silva Junior et al. 2021.
Note: PPCDAm = Plan for the Prevention and Control of Deforestation in the Legal Amazon.

BOX 2.4

The emergence of opportunistic climate institutions in India

In India, active but opportunistic climate institutions have emerged across multiple ministries because a climate lens has been layered onto established bodies. Building on existing priorities—including increasing energy supply and security, and controlling air pollution—these institutions have emerged, crucially, without strong national mitigation-centric strategic institutions, such as a climate change framework law, that could have triggered backlash due to prevailing discourses around potential trade-offs between climate and development goals.

Previous efforts to create more strategic institutions, specifically related to the National Action Plan on Climate Change, have been politically difficult. In contrast, opportunistic, sector-based, more bottom-up climate institutions—initiatives around solar energy, energy efficiency, and electric vehicles, which are delivering significant mitigation gains, primarily as co-benefits to developmental aims—have been successful. Between 2014 and 2022, India's renewable energy power generation capacity, including hydropower, increased by a factor of 2.2, with solar power growing from 3 to 63 gigawatts (Government of India 2023), thereby contributing to increased energy supply and enhanced energy security. And by 2018, energy efficiency improvements since 2010 prevented 12 percent of additional annual energy use (IEA 2021). These opportunistic climate institutions are also creating new interests in these sectors; as a result, climate is becoming mainstreamed, contributing to shifts in the political economy that could reinforce climate action going forward.

Source: Pillai and Dubash 2021.

Unstable climate institutions might not survive when the climate agenda is caught in crossfire politics—that is, when climate narratives are mitigation-centric but high levels of contestation and polarization surround the climate agenda. Climate can become a hot topic when it is embroiled in a polarized and ideologically fraught political landscape, as seen in Australia and the United States. In such cases, climate change can come to represent broader issues of political affiliation and identity politics, with a strong undertow of vested interests. Shifting narratives or layering approaches can help embed climate policies in existing programs, such as energy procurement, to protect climate institutions and policies from rollback after a change in government.

Institutions that are a good fit for the political economy can cause more enabling conditions to emerge. Climate governance institutions can trigger feedback in the rest of the political economy, with implications for the evolution of climate governance over time. For example, where mitigation-centric strategic climate institutions are prematurely established in contexts without climate consensus or where climate politics are contested, such institutions can trigger negative feedback, increasing polarization and opposition to the institution itself and climate action more broadly, as happened in Brazil. A better fit in such contexts might be opportunistic institutions, if they lead to positive feedback by supporting the emergence of pro-climate interests and mainstreaming, as seen in India. Alternatively, sectoral or subnational institutions that focus largely on existing priorities and have mitigation as a co-benefit may deliver emissions reductions and subtle shifts in the political economy in a more bottom-up manner, as happened in Australia and South Africa (boxes 2.5 and 2.6).

BOX 2.5

Political polarization undermines climate institutions in Australia

Australia's highly polarized political discourse around climate mitigation has undermined opportunities for building durable climate institutions (MacNeil 2021). The Australian economy's high resource dependency (particularly on fossil fuels) has fostered powerful vested interests in key emitting sectors that have consistently mobilized against climate action over the past three decades. Despite the establishment between 2007 and 2013 of several climate institutions—including the Australian Renewable Energy Agency, Carbon Pollution Reduction Scheme, Clean Energy Act and Clean Energy Futures Package, Clean Energy Finance Corporation, Clean Energy Regulator, Climate Change Authority, Climate Commission, and Multi-Party Committee on Climate Change—most were rolled back or undermined after a change in government. In 2014, the government disbanded the Climate Commission and repealed the carbon price legislated under the Clean Energy Act.

Given the high levels of national-level political contestation and polarization, subnational climate institutions have become the primary driver of policy-based emissions reductions in Australia, enabled to a degree by national interventions that improved investment conditions and incentives around renewable energy. Specifically, the Australian Renewable Energy Agency and Clean Energy Finance Corporation have been essential in providing "pull" instruments in the form of government funding and job creation in renewable energy. Whether these or similar interventions are enough to shift interests and other political economy factors to more strategic climate institutional action is yet to be seen.

Following the most extensive bushfires in the country's history, in the summer of 2019–20, and major floods in early 2022, a new government came to power in May 2022 with a mandate to strengthen Australia's response to climate change. The government passed a climate change framework law enshrining a 43 percent reduction in emissions by 2030, a 2050 net zero target, and an enhanced role for independent expert advice. Separate legislation was passed to strengthen the "safeguard mechanism" to ratchet down emissions from major sources.

BOX 2.6

Why coal politics trump climate institutions in South Africa

Climate institutions in South Africa have struggled to take hold because of highly contested politics in fossil fuel-based sectors. Coal dominates the power mix (figure B2.6.1) and accounts for 84 percent of emissions in the energy sector; vested interests in this and other coal-dominated sectors—including electricity, synthetic fuel, and steel—are deeply embedded and powerful. In contrast to other cases, where cross-sectoral coordination is viewed as the key challenge, South Africa's main climate challenge lies in reorienting a single sector dominated by a few powerful actors (Baker et al. 2015). Climate institutions outside the energy sector, such as the 2011 National Climate Change Response White Paper, and interventions within the sector, such as the Integrated Resource Plan and the Renewable Energy Independent Power Producers Procurement Programme, have been delayed or stalled, or have had limited effect.

More recently, efforts have shifted to strategic climate institutions centering on a just transition narrative, through the Just Transition Framework (Presidential Climate Commission 2022), which speaks to social and political priorities relating to redistributive development and addresses concerns of some powerful interest groups (namely labor unions), with mitigation as a co-benefit. At the same time, the energy supply crisis and rapid performance decline in South Africa's aging coal infrastructure have opened the way for reforms that might allow for the needed reorientation of the energy sector. Ultimately, these factors will likely drive the bulk of mitigation in South Africa. Additionally, discussions increasingly center around the European Union's Carbon Border Adjustment Mechanism and its implications for the South African export industry. To preserve industrial competitiveness and increase global climate action, Carbon Border

(Continued)

BOX 2.6

Why coal politics trump climate institutions in South Africa (continued)

Adjustment Mechanisms aim to balance the prices of goods produced in a jurisdiction with carbon pricing mechanisms and externally produced goods by applying an equivalent carbon price on imported goods. Possibly affecting a large share of South African producers—including those in the steel industry, which employs 28,000 people—such a mechanism will have a great impact on the country's political economy of climate action (Presidential Climate Commission 2023). With highly interdependent national, provincial, and local spheres of government, generating a climate consensus political narrative and building strong government coordination mechanisms and capabilities could help open the way for more strategic climate institutions, accounting for the responsibility of state actors and smoothing the transition.

FIGURE B2.6.1. **South Africa's power mix, 1990–2021**

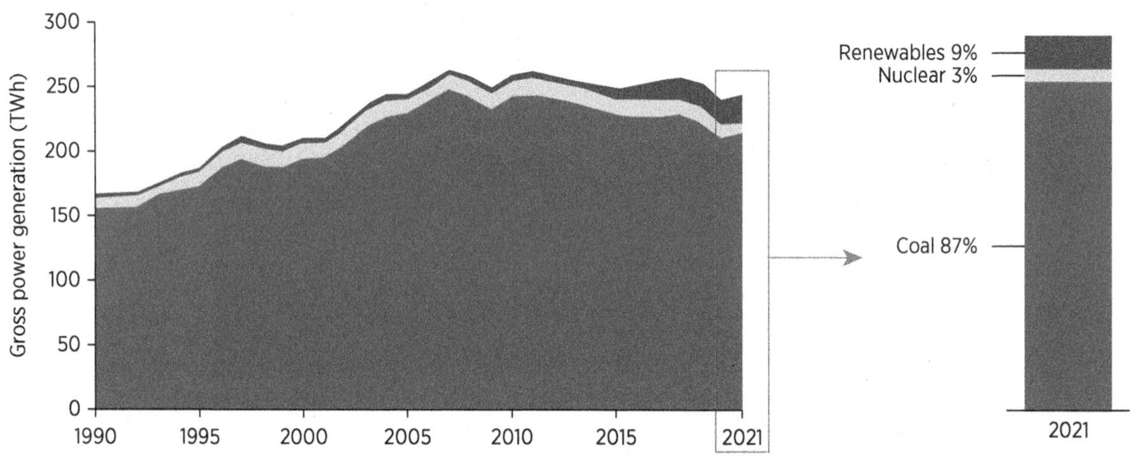

Source: Climate Transparency 2022.
Note: TWh = terrawatt-hours.

Source: Tyler and Hochstetler 2021.

From opportunistic or unstable to strategic climate institutions

At first, opportunistic or unstable climate institutions may be the only feasible form in the political economy. In such cases, governments can work within the existing political economy by layering climate governance functions into existing institutional structures, embedding climate into other political priorities with a focus on "win-win" or synergies, and frontloading initiatives that help bring down costs, boost innovation, and build capacity. Building climate into existing institutions—instead of waiting for the perfect conditions—can enable quicker and more ambitious action. To this end, governments can adopt a range of approaches, such as

- *Co-benefits and development synergies*, by mainstreaming climate through existing high-level political priorities, emphasizing co-benefits and synergies, or focusing on topical issues, such as energy access or job creation
- *Integration*, by building climate objectives or mandates into sectoral institutions (for example, adding a renewable energy contingent to an energy planning and procurement body or an electric vehicle unit to a transportation authority)
- *Pilot programs*, by using experimental approaches to allow for institutional learning, proof of concept, and litmus for social and political adjustment

- *Public investments*, by using institutions to provide green public investment and research and development strategies to drive down costs, crowd in other finance sources, spur innovation, and support emergent green interests
- *Monitoring*, by building capacity to provide information that can later be used in enforcement
- *Capacity building*, through education and training programs across government departments, industry, and professional groups, and through formal education.

Governments should aim to move from opportunistic or unstable to enduring, strategic climate institutions. Countries cannot achieve transformative systemic change without widespread social buy-in because of the changes to production, consumption, governance, and lifestyles such transformation implies. To develop buy-in, governments will need to develop strategic climate institutions that show how to reach society's goals alongside or through climate goals and create a desirable vision for the future that aligns with people's deeper values, principles, and aspirations. By creating more enabling conditions, governments can lay the groundwork for and begin developing strategic institutions. For example, they can start to establish climate governance institutions that help mediate interest groups and build consensus around narrative and high-level direction setting, facilitate and inform stakeholder engagement and alignment, foster supportive coalitions, and improve the overall institutional context.

Strategic climate institutions: Framework legislation, long-term strategies, and just transition frameworks

Climate change framework law, long-term strategies (LTSs), or just transition frameworks can provide an overarching institutional basis for making and implementing climate policy. In contrast to opportunistic or unstable institutions, which often target a single sector or issue, these strategic institutions provide a whole-of-economy framework that integrates multiple objectives and existing institutions. They provide a shared vision and mandate for climate policies and action built around a holistic approach that aligns climate targets with economic development goals, social objectives, and citizen engagement imperatives.

Climate change framework legislation

Climate change framework legislation provides a legal basis for climate policy. Over 60 countries have adopted framework legislation to tackle climate change, and more are developing or considering it. Such laws can help countries design their own effective and comprehensive strategic climate legislation, from setting targets and developing strategies to engaging and overseeing stakeholders. But they also provide new avenues for stakeholders to influence climate action—for instance, when they create new coordination bodies, improve and structure stakeholder engagement, or translate nonbinding ambitions into binding targets and open the door to using litigation as a commitment device (box 2.7).

The comprehensiveness of climate change framework legislation varies across countries, representing different stages of political economy readiness. Some countries may need to implement alternative governance interventions before or instead of adopting comprehensive framework legislation, or to develop more limited framework legislation that they can enhance at a later stage. Several countries have taken the latter approach, including only some of the elements outlined in table 2.3 in their national climate legislation. For example, the figure shows that in 2023, of the 33 economies identified in the

BOX 2.7

Climate change litigation as a tool for improved climate action

Established legal frameworks help citizens hold governments to account for their actions and inactions and serve as a binding regulatory structure when political administrations and priorities change. These frameworks represent a way of institutionalizing and protecting defined priorities over time: governments are obliged to adhere to—and can be prosecuted for violating—their own laws and regulations. Such frameworks can also serve to make governments' commitments more credible in the long term, which is necessary to influence household and business decisions. Climate and environmental activists and nongovernmental organizations also increasingly use climate litigation to hold governments and companies legally responsible for contributing to global warming (Schiermeier 2021).

Within the last 20 years, the number of lawsuits related to climate change has grown from less than 10 in the early 2000s to nearly 200 in 2020 (figure B2.7.1). Today, more than 2,000 cases of climate change litigation have been identified worldwide, with cases filed in the United States making up the largest share (71 percent), followed by Australia (6 percent), the United Kingdom (4 percent), and the European Union (3 percent); but numbers are also growing in the global South (Setzer and Higham 2022).

Most cases have been filed by nongovernmental organizations, individuals, or both acting together; and governments remain the most frequent targets in climate litigation (Setzer and Higham 2022). For example, in 2015, the Dutch environmental group Urgenda Foundation, together with 900 Dutch citizens, sued the Dutch government for failing to take enough action to prevent climate change and adhere to the agreement of keeping global temperature increases within 2°C of preindustrial conditions. The court in The Hague rebuked the government for its pledge to reduce emissions to only 17 percent below 1990 levels by 2020 as inadequate and ordered it to increase this number to at least 25 percent. The court's decision constitutes "the first decision by any court in the world ordering states to limit greenhouse gas emissions for reasons other than statutory mandates."[a] Similarly, supported by the establishment of national judicial climate change agenda implementation mechanisms, the Brazilian public increasingly uses litigation to promote climate goals and has had several climate-related appeals before the federal supreme court. For example, in 2020, four political parties filed a case before the court, denouncing the federal government's alleged failure to adopt administrative measures concerning the Amazon Fund; the case resulted in reactivation of the fund in 2023 under the new administration.[b]

Although cases filed by nongovernmental organizations and individuals make up the largest share of climate litigation, governments, companies, and trade associations can also file climate cases in the courts, as illustrated by multiple examples from the United States (Setzer and Higham 2022). In 2023, Multnomah

FIGURE B2.7.1. Climate change-related lawsuits, 2000–21

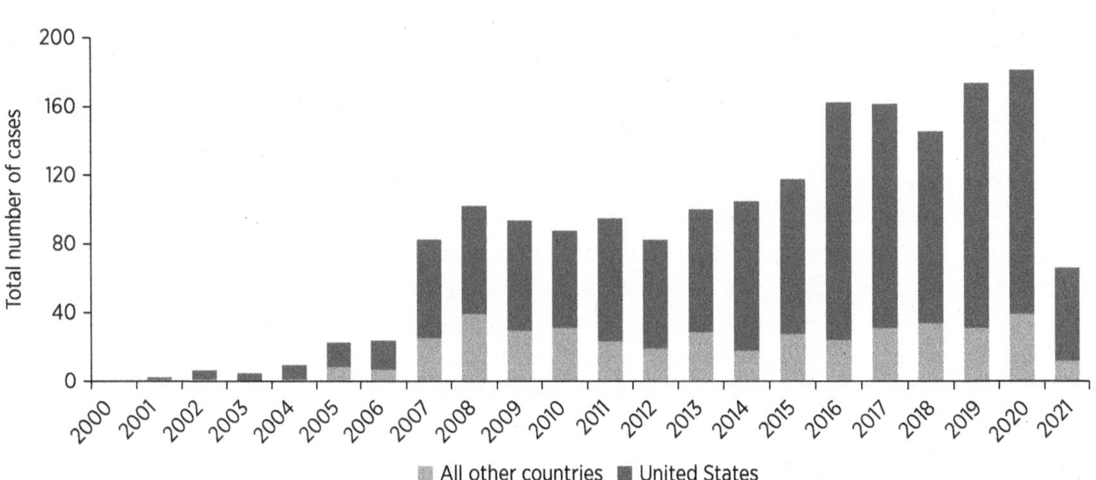

Source: Schiermeier 2021.

(Continued)

BOX 2.7

Climate change litigation as a tool for improved climate action (continued)

County, Oregon, sued Exxon, Chevron, and other fossil fuel companies and industry groups for over US$50 billion to reduce and mitigate the harms caused by climate change. The county also accused consulting company McKinsey of supporting the industry in selling fossil fuel products and in falsely promoting them as harmless to the environment (Mindrock 2023).[c] The lawsuit follows similar cases filed against oil companies in previous years by various US cities.[d]

a. Global Climate Change Litigation database, Urgenda Foundation v. State of the Netherlands, https://climatecasechart.com/non-us-case/urgenda-foundation-v-kingdom-of-the-netherlands/.
b. Global Climate Change Litigation database, PSB et al. v. Brazil (on Amazon Fund), https://climatecasechart.com/non-us-case/psb-et-al-v-brazil/.
c. U.S. Climate Change Litigation database, County of Multnomah v. Exxon Mobil Corp. https://climatecasechart.com/case/county-of-multnomah-v-exxon-mobil-corp/.
d. U.S. Climate Change Litigation database, City of Oakland v. BP p.l.c. https://climatecasechart.com/case/people-state-california-v-bp-plc-oakland/; City of Charleston v. Brabham Oil Co. https://climatecasechart.com/case/city-of-charleston-v-brabham-oil-co/; City of New York v. Exxon Mobil Corp. https://climatecasechart.com/case/city-of-new-york-v-exxon-mobil-corp/.

TABLE 2.3. **A comparison of climate change framework legislation across 33 economies**

	Long-term targets	Midterm and sectoral	Risk and vulnerability assessments	Climate change strategies and plans	Policy instruments	Independent expert advice	Mechanism for coordination	Stakeholder engagement	Subnational governments	Financing implementation	Monitoring and review of progress	Oversight and accountability
Austria												
Benin												
Brazil												
Bulgaria												
Colombia												
Croatia[a]												
Denmark												
Finland												
France												
France[a]												
Germany												
Guatemala												
Honduras												
Ireland												

(Continued)

TABLE 2.3. A comparison of climate change framework legislation across 33 economies (continued)

	Long-term targets	Midterm and sectoral	Risk and vulnerability assessments	Climate change strategies and plans	Policy instruments	Independent expert advice	Mechanism for coordination	Stakeholder engagement	Subnational governments	Financing implementation	Monitoring and review of progress	Oversight and accountablity
Japan (adaptation)	Not integrated	Not integrated	Not integrated	Somewhat integrated	Somewhat integrated	Somewhat integrated	Somewhat integrated	Somewhat integrated	Somewhat integrated	Not integrated	Somewhat integrated	Not integrated
Japan (mitigation)	Not integrated	Not integrated	Somewhat integrated	Somewhat integrated	Somewhat integrated	Somewhat integrated	Somewhat integrated	Somewhat integrated	Somewhat integrated	Not integrated	Integrated	Not integrated
Kenya	Not integrated	Not integrated	Somewhat integrated	Somewhat integrated	Somewhat integrated	Not integrated	Integrated	Integrated	Integrated	Not integrated	Somewhat integrated	Somewhat integrated
Korea, Rep.[a]	Somewhat integrated	Somewhat integrated	Integrated	Integrated	Integrated	Not integrated	Integrated	Integrated	Integrated	Not integrated	Integrated	Integrated
Liechtenstein	Not integrated	Not integrated	Not integrated	Not integrated	Somewhat integrated	Somewhat integrated	Not integrated	Not integrated	Not integrated	Not integrated	Not integrated	Not integrated
Malta	Not integrated	Not integrated	Not integrated	Somewhat integrated	Somewhat integrated	Not integrated	Somewhat integrated	Somewhat integrated	Not integrated	Not integrated	Integrated	Somewhat integrated
Mexico	Integrated	Integrated	Integrated	Integrated	Integrated	Not integrated	Integrated	Not integrated	Integrated	Not integrated	Integrated	Not integrated
Micronesia, Fed. Sts.	Not integrated	Not integrated	Not integrated	Somewhat integrated	Not integrated	Somewhat integrated	Somewhat integrated	Not integrated	Not integrated	Somewhat integrated	Integrated	Not integrated
Netherlands	Somewhat integrated	Somewhat integrated	Not integrated	Somewhat integrated	Somewhat integrated	Not integrated	Not integrated	Not integrated	Not integrated	Not integrated	Somewhat integrated	Somewhat integrated
New Zealand	Somewhat integrated	Somewhat integrated	Integrated	Not integrated	Integrated	Integrated	Somewhat integrated	Not integrated	Not integrated	Not integrated	Integrated	Not integrated
Norway	Somewhat integrated	Somewhat integrated	Not integrated	Somewhat integrated	Somewhat integrated	Somewhat integrated	Somewhat integrated	Not integrated	Not integrated	Somewhat integrated	Somewhat integrated	Somewhat integrated
Pakistan	Not integrated	Not integrated	Integrated	Integrated	Somewhat integrated	Somewhat integrated	Integrated	Somewhat integrated	Somewhat integrated	Somewhat integrated	Somewhat integrated	Somewhat integrated
Papua New Guinea	Not integrated	Not integrated	Not integrated	Somewhat integrated	Integrated	Not integrated	Not integrated	Not integrated	Somewhat integrated	Not integrated	Integrated	Not integrated
Paraguay	Not integrated	Not integrated	Not integrated	Integrated	Not integrated	Not integrated	Integrated	Integrated	Not integrated	Somewhat integrated	Somewhat integrated	Not integrated
Peru	Not integrated	Not integrated	Not integrated	Integrated	Somewhat integrated	Somewhat integrated	Integrated	Somewhat integrated	Somewhat integrated	Not integrated	Integrated	Not integrated
Philippines	Integrated	Integrated	Integrated	Integrated	Integrated	Integrated	Integrated	Integrated	Somewhat integrated	Somewhat integrated	Somewhat integrated	Integrated
Slovenia	Not integrated	Not integrated	Integrated	Somewhat integrated	Integrated	Somewhat integrated	Somewhat integrated	Integrated	Integrated	Not integrated	Integrated	Somewhat integrated
Sweden	Not integrated	Not integrated	Not integrated	Somewhat integrated	Not integrated	Not integrated	Not integrated	Not integrated	Not integrated	Not integrated	Somewhat integrated	Somewhat integrated
Switzerland[a]	Somewhat integrated	Integrated	Not integrated	Not integrated	Somewhat integrated	Not integrated	Not integrated	Not integrated	Somewhat integrated	Somewhat integrated	Not integrated	Somewhat integrated
Taiwan, China	Somewhat integrated	Somewhat integrated	Somewhat integrated	Somewhat integrated	Somewhat integrated	Somewhat integrated	Not integrated	Somewhat integrated	Integrated	Not integrated	Integrated	Not integrated
United Kingdom	Integrated	Integrated	Integrated	Integrated	Integrated	Integrated	Integrated	Integrated	Somewhat integrated	Somewhat integrated	Somewhat integrated	Integrated

● Integrated ● Somewhat integrated ● Not integrated

Source: Updated version of figure A.2 in World Bank 2020.
Note: The rating of EU member countries on their climate change framework legislation considers their national policies only. Even if a member country does not have, for example, a long-term target in its national legislation, it is still covered by EU frameworks, such as Fit For 55 and others.
a. Updated law.

"Reference Guide to Climate Change Framework Legislation" (World Bank 2020), only 12 have adopted long-term targets, 13 do not define roles for subnational governments, 12 do not integrate stakeholder engagement, and 17 do not include provisions for financing for implementation.

Gaps in climate change framework legislation can undermine its effectiveness by allowing political economy constraints to persist or reemerge. For example, without legally binding targets or rules relating to ambition ratcheting to keep legal targets in line with climate goals, climate change legislation can become less effective over time, opening the way for backsliding, as happened in Mexico (box 2.8). Similarly, the failure to establish organizations responsible for implementation—or clear rules about delegating roles and responsibilities and financing mechanisms—can make the law less credible or more difficult to enforce. For countries that begin with a more limited climate framework law, planning and enacting enhancements to strengthen its effectiveness and durability are vital.

Long-term strategies

LTSs offer a complementary approach to developing a climate governance framework, which can be based on or inform climate framework laws.[1] As of September 2023,

---

BOX 2.8
Mexico's General Law on Climate Change

In 2012, Mexico became the first large oil-producing emerging economy to adopt climate legislation when its Parliament passed the General Law on Climate Change (Ley General de Cambio Climático, or LGCC). The LGCC established an aspirational goal of reducing emissions by 30 percent below an unspecified baseline scenario by 2020, and by 50 percent below the 2000 emission level by 2050.

The law's key impacts included establishing key federal-level institutions to deal with climate change, defining responsibilities for states and municipalities, and defining long-term objectives. It effectively set the basis for climate policy in Mexico, including the National Strategy on Climate Change and Special Program on Climate Change (Averchenkova and Guzman Luna 2018).

In April 2018, Mexico amended the LGCC to align it with the Paris Agreement. This amendment included revisions to greenhouse gas and black carbon targets, indicating that emissions would peak by 2026 and that the country would reduce greenhouse gas intensity per unit of gross domestic product by about 40 percent between 2013 and 2030.

Although the Special Program on Climate Change had a mixed implementation record between 2014 and 2018, achieving only 43 percent of the set goals (INECC 2017), the LGCC has played a major role in guiding the low-carbon transition in the energy sector. The Energy Transition Law, adopted in 2015, builds on emission targets set in the LGCC and has helped drive the development of renewable energy in the country. It sets targets for clean energy generation (35 percent by 2024 and 50 percent by 2050), enabling mechanisms for renewables, such as long- and medium-term electricity auctions and the Clean Energy Certificates market. As a result, wind and solar generation tripled over the five years leading to 2021 (Gabbatiss 2021).

Since the end of 2018, however, the country has seen a change in political commitment toward climate change and renewable energy policies (Parish Flannery 2021). Delays in revising policies under the Special Program on Climate Change, structural and operational changes to the climate change fund and its continued difficulties leveraging resources, budget cuts for climate change activities, and increased support to fossil fuels (ICM 2020) threaten the country's ability to meet the objectives of the law and implement its nationally determined contribution. The overall emission reduction objectives set in the law stand, however, and changing them would require an agreement of the legislature.

67 countries have LTSs under the United Nations Framework Convention on Climate Change, providing a realistic pathway toward long-term objectives and identifying useful milestones for shorter-term strategies and plans. Despite the possibility of reducing emissions by 10 or 20 percent by acting only on the emissions that are cheapest to abate, a transition toward net zero requires action on all emissions sources and thus a different approach to sequencing and prioritization. Rather than identifying the cheapest emissions to abate, achieving net zero emissions requires designing the least-cost transition for each sector and emission source, with the right sequencing and timing of actions. LTSs can be the basis for developing short-term climate plans or strategies, including nationally determined contributions.

LTSs have multiple roles and functions in climate governance, starting with informing political debates and choices. If the institutional setup is well designed and the process inclusive and technically sound, design of an LTS can inform political debates in countries and identify the critical choices and milestones they need to make. It can also sometimes inform the choice of a long-term target—for example, the date to achieve net zero emissions. But this requires designing LTSs in an iterative manner, using the process to capture knowledge from public and private actors, and offering opportunities for all stakeholders to provide feedback and contribute to the discussion. The original LTS effort carried out in France in 2012 to elaborate the low-carbon national strategy, or *Stratégie Nationale Bas Carbone*, prioritized the generation of whole-of-economy pathways consistent with keeping global warming under 2°C and invited different stakeholder groups and experts to develop their own scenarios. The strong central team, comprising government officials and hired specialists with modeling and technical expertise, played a crucial role; that team collated and analyzed a wide range of technical scenarios from different stakeholder groups, and used them to create four overarching alternative visions or pathways, which informed the political debate. The team's ability to incorporate input and experts from diverse sectors, and to guide technical deliberations and synthesis, gave these four pathways credibility, and they were accepted as a fair range of options for national consideration. Similarly, Costa Rica's LTS, the *Plan Nacional de Descarbonización*, used a whole-of-economy pathway that presented targets and timelines to all emitting sectors to enable technical discussions that could explore the extent of necessary changes and frame existing barriers and the required enabling conditions (see World Bank 2023).

When an LTS receives widespread support or is embedded in law, it can also provide a powerful instrument to maintain momentum, coordinate action across sectors, and offer a benchmark to measure progress over time. The need for a coordinated transformation—for example, between the power and transportation sectors—is a well-known obstacle to decarbonization. An LTS can provide the key milestones to support such coordination—for example, with indicative targets for the share of electric vehicles on the roads, which both the energy and transportation ministries can use to design their policies and plans (Fay et al. 2015). LTSs can also provide a set of milestones, such as the share of renewables in the power mix, the modal share of rail in freight, or the number of retrofitted dwellings. Ministries and public agencies can then use these milestones to set up monitoring and evaluation systems to track progress and identify lagging sectors that require additional interventions; private sector stakeholders and the general public can use them in their own decision-making and to assess the government's performance. For this to happen, it is important to make sure that the LTS's objectives become part of the functional mandates of various ministries and agencies. In France, the Climate Change Law of 2015 has empowered the development and implementation of the LTS. This implementation notably includes the multiyear energy program, or *Programmation Pluriannuelle de l'Energie*, which establishes the government's energy priorities over 10 years. Costa Rica made efforts to build LTSs into both national law and international cooperation deals—for example, agreeing to policy-based loans with development partners that align financing on favorable terms with achieving environmental policy milestones. In doing so, it strengthened the interest of the Ministry of Finance and investor groups in ensuring the targets were met.

Just transition principles and frameworks

A growing number of countries are developing just transition frameworks, helping those countries establish a social mandate and guiding principles for climate action based on equity and fairness. These frameworks offer a people-first approach that considers how to include and support people and societies to enhance the equity and fairness of climate policy decisions and outcomes. A notable example is South Africa's Just Transition Framework, adopted by the cabinet in 2022 (box 2.9). Just transitions

BOX 2.9

South Africa's Just Transition Framework

South Africa is regarded as an international leader when it comes to the ambition of its commitments on climate mitigation, specifically among developing countries. This position is especially remarkable given the country's highly emissions-intensive economy. It is Africa's largest greenhouse gas emitter and the world's twelfth largest, and at least 84 percent of its carbon dioxide emissions come from the coal industry (Ritchie, Roser, and Rosado 2020). As well as submitting two nationally determined contributions (in 2016 and 2021), its Low-Emission Development Strategy (2020), and its National Adaptation Plan (2020) to the United Nations Framework Convention on Climate Change, the country has established several climate-related laws, policies, and strategies. They include the White Paper on the Promotion of Renewable Energy and Clean Energy Development (2003), National Climate Change Response White Paper (2011), Green Transport Strategy (2018), National Energy Efficiency Strategy (2019), Carbon Tax Act (2019), and Green Finance Taxonomy (2022).

Climate action in South Africa faces both challenges and opportunities because the political economy surrounding it remains a contentious issue and slows progress. South Africa's distinctive cabinet-approved Just Transition Framework not only exemplifies this complexity but also provides a positive perspective. For example, although the extensive consultations required to build consensus contributed to the delay in passing the Climate Change Bill, they also highlight the commitment to a just and equitable transition and underpin the government's responsibility in securing "ecologically sustainable development and use of natural resources while promoting justifiable economic and social development," as stated in the country's 1996 Constitution. Once it becomes law, the Climate Change Bill will not only enhance the overall governance framework for climate change response but also formally codify South Africa's nationally determined contribution targets, further emphasizing the nation's commitment to addressing climate change.

Neverthless, after some initial success, South Africa's pioneering Renewable Energy Independent Power Procurement Program was undermined by influential actors with vested interests in the coal industry and ideological oppositions to private sector involvement in the energy sector. Those actors included the state monopoly power company Eskom, industry trade unions, and the Ministry of Energy and Minerals. Opposition to the energy transition also stems from and is amplified by concerns about justice in a country still marked by extreme inequality and exclusion 30 years after the end of Apartheid. In particular, South Africa's powerful trade unions have concerns about labor and the sectoral and spatial effects of the transition.

(Continued)

BOX 2.9

South Africa's Just Transition Framework (continued)

Uncertainty regarding the equitable distribution of benefits from the transition toward renewables increases fear and resistance (World Bank Group 2022). Altogether, creating a shared vision for the climate transition with support from important actors, including citizens, has been a major challenge.

To address this situation, the country established the Presidential Climate Commission in 2020, bringing together representatives from government, business, labor, civil society, and research and academic institutions to coordinate and oversee a just transition toward a low-carbon, inclusive, climate-resilient economy and society. As one of its first tasks, it developed a Just Transition Framework that meets the needs of all social partners. Adopted by the cabinet in 2022, the framework supports South Africa's broader efforts to redesign the economy to the benefit of most citizens and enable deep, just, and transformational shifts in the context of delivering an effective response to climate change. It is built around three principles:

1. *Distributive justice* distributes risks and opportunities fairly, cognizant of gender, race, and class inequalities.
2. *Restorative justice* addresses historical damages against communities.
3. *Procedural justice* empowers and supports workers and communities through the transition.

The framework spotlights at-risk sectors and value chains and lays out key policy areas for a just transition, governance imperatives, and financing.

FIGURE 2.2. **Four principles for a just transition**

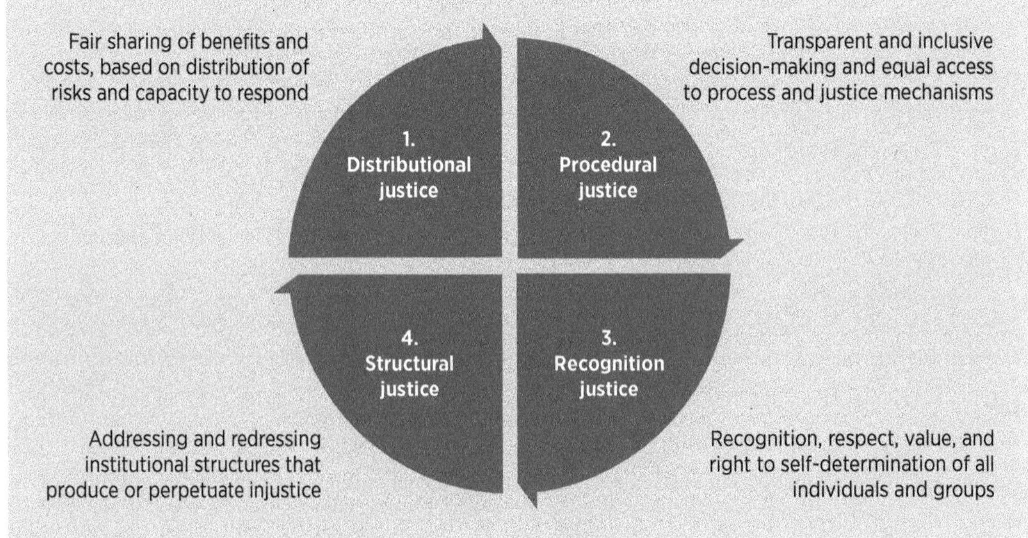

Source: Original figure developed for this report.

will vary depending on local context and as defined through local processes, but they have four common guiding principles: distributional justice, procedural justice, recognition justice, and structural justice (figure 2.2). When designing just transition frameworks to guide their own journey to net zero, governments can draw lessons from other countries' experiences (Krawchenko and Gordon 2021). They can also learn from and build on existing social service and social justice architectures, including newer and more innovative instruments.

Countries can also integrate just transition principles into other strategies and laws, including nationally determined contributions (UNFCCC Secretariat 2021). For example, the European Commission's proposed Fit for 55 legislative package includes a Just Transition Mechanism and Social Climate Fund. Industry- or sector-specific strategies that include just transition principles and mechanisms include Canada's Task Force on the Just Transition for Canadian Coal Power Workers, Spain's Just Transition Agreements, and New Zealand's Just Transition Unit (Krawchenko and Gordon 2021).

Note

1. This section is based on World Bank and IDDRI, forthcoming (background paper for this book).

References

Averchenkova, A., S. Fankhauser, and J. J. Finnegan. 2021. "The Impact of Strategic Climate Legislation: Evidence from Expert Interviews on the UK Climate Change Act." *Climate Policy* 21(2): 251–63. https://doi.org/10.1080/14693062.2020.1819190.

Averchenkova, A., and S. L. Guzman Luna. 2018. "Mexico's General Law on Climate Change: Key Achievements and Challenges Ahead." Grantham Research Institute on Climate Change and the Environment and Centre for Climate Change Economics and Policy, London School of Economics and Political Science, London.

Baker, L., J. Burtion, C. Godinho, and H. Trollip. 2015. "The Political Economy of Decarbonisation: Exploring the Dynamics of South Africa's Electricity Sector." Research Report Series, Energy Research Centre, University of Cape Town, Cape Town. https://doi.org/10.13140/RG.2.1.4064.9040.

Climate Transparency. 2022. "South Africa." Climate Transparency Report 2022. https://www.climate-transparency.org/wp-content/uploads/2022/10/CT2022-South-Africa-Web.pdf.

Dubash, N. K. 2021. "Varieties of Climate Governance: The Emergence and Functioning of Climate Institutions." *Environmental Politics* 30 (1): 1–25. https://doi.org/10.1080/09644016.2021.1979775.

Dubash, N., A. V. Pillai, C. Flachsland, K. Harrison, K. Hochstetler, M. Lockwood, R. Macneil, M. Mildenberger, M. Paterson, F. Teng, and E. Tyler. 2021. "National Climate Institutions Complement Targets and Policies." *Science* 374 (6568): 690–93.

Fay, M., S. Hallegatte, A. Vogt-Schilb, J. Rozenberg, U. Narloch, and T. Kerr. 2015. *Decarbonizing Development: Three Steps to a Zero-Carbon Future.* Washington, DC: World Bank. http://hdl.handle.net/10986/21842.

Gabbatiss, J. 2021. "The Carbon Brief Profile: Mexico." Carbon Brief Country Profiles, June 4, 2021. https://www.carbonbrief.org/the-carbon-brief-profile-mexico.

Government of India. 2023. "Annual Report 2022-2023." https://mnre.gov.in/annual-reports-2022-23.

Grantham Research Institute (Grantham Research Institute on Climate Change and the Environment). 2022. "Framework Law on Climate Change – Chile." Climate Change Laws of the World, Grantham Research Institute, London. https://climate-laws.org/document/framework-law-on-climate-change-chile_dc8a.

Hanusch, M., ed. 2023. *A Balancing Act for Brazil's States of the Brazilian Amazon: An Economic Memorandum.* Washington, DC: World Bank.

Hochstetler, K. 2021. "Climate Institutions in Brazil: Three Decades of Building and Dismantling Climate Capacity." *Environmental Politics* 30 (1): 49–70. https://doi.org/10.1080/09644016.2021.1957614.

ICM (Iniciativa Climatica de Mexico). 2020. "México Sin Ambición Para Atender la Crisis Climática." ICM, December 16, 2020. https://www.iniciativaclimatica.org/mexico-sin-ambicion-para-atender-la-crisis-climatica/.

IEA (International Energy Agency). 2021. "E4 Country Profile: Energy Efficiency in India." IEA, Paris. https://www.iea.org/articles/e4-country-profile-energy-efficiency-in-india.

INECC (Instituto Nacional de Ecología y Cambio Climático, Mexico). 2017. *Evaluación Estratégica del Programa Especial de Cambio Climático 2014–2018.* Mexico City: INECC. https://cambioclimatico.gob.mx/evaluacion-estrategica-del-programa-especial-de-cambio-climatico-2014-2018-2017/.

Krawchenko, T. A., and M. Gordon. 2021. "How Do We Manage a Just Transition? A Comparative Review of National and Regional Just Transition Initiatives." *Sustainability* 13 (11): 6070. https://doi.org/10.3390/su13116070.

Lockwood, M. 2021. "A Hard Act to Follow? The Evolution and Performance of UK Climate Governance." *Environmental Politics* 30 (1): 26–48. https://doi.org/10.1080/09644016.2021.1910434.

MacNeil, R. 2021. "Swimming against the Current: Australian Climate Institutions and the Politics of Polarisation." *Environmental Politics* 30 (1): 162–83. https://doi.org/10.1080/09644016.2021.1905394.

Mindrock, C. 2023. "US Climate Change Lawsuit Seeks $50 Billion, Citing 2021 Heat Wave." *Reuters*, June 22, 2023. https://www.reuters.com/world/us/us-climate-change-lawsuit-seeks-50-billion-citing-2021-heat-wave-2023-06-22/.

Parish Flannery, N. 2021. "Political Risk Analysis: Is Mexico Declaring War against Clean Energy?" *Forbes*, April 22, 2021. https://www.forbes.com/sites/nathanielparishflannery/2021/04/22/political-risk-analysis-is-mexico-declaring-war-against-clean-energy/?sh=d35b182701a5.

Pillai, V. A., and N. K. Dubash. 2021. "The Limits of Opportunism: The Uneven Emergence of Climate Institutions in India." *Environmental Politics* 30 (sup1): 93–117. https://doi.org/10.1080/09644016.2021.1933800.

Presidential Climate Commission. 2022. "A Framework for a Just Transition in South Africa." Republic of South Africa. https://www.climatecommission.org.za/just-transition-framework.

Presidential Climate Commission. 2023. "Carbon Border Adjustment Mechanisms and Implications for South Africa." A Presidential Climate Commission Working Paper, Republic of South Africa. https://www.climatecommission.org.za/publications/cbam.

Ritchie, H., M. Roser, and P. Rosado. 2020. "CO_2 and Greenhouse Gas Emissions." Our World in Data, August 2020. https://ourworldindata.org/co2-and-greenhouse-gas-emissions.

Schiermeier, Q. 2021. "Climate Science is Supporting Lawsuits that Could Help Save the World." *Nature* 597: 169. https://www.nature.com/articles/d41586-021-02424-7.

Setzer, J., and C. Higham. 2022. *Global Trends in Climate Litigation: 2021 Snapshot*. London: Grantham Research Institute on Climate Change and the Environment. https://eprints.lse.ac.uk/117652/.

Silva Junior, C. H. L., A. C. M. Pessôa, N. S. Carvalho, J. B. C. Reis, L. O. Anderson, and L. E. O. C. Aragão. 2021. "The Brazilian Amazon Deforestation Rate in 2020 Is the Greatest of the Decade." *Nature Ecology and Evolution* 5 (2): 144–45. https://doi.org/10.1038/s41559-020-01368-x.

Sridhar, A., N. K. Dubash, A. Averchenkova, C. Higham, O. Rumble, and A. Gilder. 2022. *Climate Governance Functions: Towards Context-Specific Climate Laws*. Grantham Research Institute on Climate Change and the Environment, London. https://www.lse.ac.uk/granthaminstitute/publication/climate-governance-functions-towards-context-specific-climate-laws/.

Tyler, E., and K. Hochstetler. 2021. "Institutionalising Decarbonisation in South Africa: Navigating Climate Mitigation and Socio-Economic Transformation." *Environmental Politics* 30 (sup1): 184–205. https://doi.org/10.1080/09644016.2021.1947635.

UNFCCC Secretariat. 2021. "Nationally Determined Contributions under the Paris Agreement. Revised Note by the Secretariat." FCCC/PA/CMA/2021/8/Rev.1, United Nations Framework Convention on Climate Change. https://unfccc.int/documents/307628.

World Bank. 2017. *World Development Report 2017: Governance and the Law*. Washington, DC: World Bank. https://www.worldbank.org/en/publication/wdr2017.

World Bank. 2020. "World Bank Reference Guide to Climate Change Framework Legislation." EFI Insight–Governance, World Bank, Washington, DC. http://hdl.handle.net/10986/34972.

World Bank. 2021. *Climate Change Institutional Assessment*. Equitable Growth, Finance and Institutions Notes – Governance. Washington, DC: World Bank. http://hdl.handle.net/10986/35438.

World Bank. 2023. *Reality Check: Lessons from 25 Policies Advancing a Low-Carbon Future*. Climate Change and Development Series. Washington, DC: World Bank. http://hdl.handle.net/10986/40262.

World Bank and IDDRI. Forthcoming. "Integrating LTS into National Decision-Making Processes." Background paper for this report.

World Bank Group. 2022. "South Africa Country Climate and Development Report." Country Climate and Development Report, World Bank, Washington, DC. http://hdl.handle.net/10986/38216.

3 Policy Sequencing

Balancing Feasibility and Long-Term Ambition

KEY INSIGHTS

Policies need to be selected for dynamic, rather than static, efficiency. The lowest-cost option today may lead to political backlash and create higher costs in the future, whereas choosing a more expensive policy today might be more dynamically efficient if it shifts the political economy to make it easier to implement more efficient policies later.

The Climate Policy Feasibility Frontier can help identify the most promising policies in a given context. Countries can strategically select and sequence policies that help reduce or overcome political economy obstacles, by building institutional capacity or creating winners who will support further policy action. Policies like institution building, which have limited direct impact on emissions, can remove barriers to decarbonization or build essential capacity in the country.

Governments can also leverage reinforcing policy feedback processes and target tipping points in the broader socio-technical-political system to accelerate transformational change. These tipping points, whether technological, social and behavioral, or political, are key to accelerating decarbonization. Strategically selecting and sequencing feasible policies does not mean climate progress will be slow.

Policy makers face hard choices between focusing on low-hanging fruits or investing in more challenging, but more transformational, strategies and policies. On the one hand, always choosing easy policies would ensure action but is unlikely to trigger the systemic changes needed to reduce emissions to zero, at least in a time frame consistent with global objectives. On the other hand, always choosing transformational policies may lead to inaction as political forces or lack of capacity render enactment, implementation, or enforcement of climate policies impossible. It is well known that the sum of least-cost marginal abatement options will not achieve the least-cost transformation needed to achieve large emission reduction, and that emission reduction options should be sequenced according to a long-term strategy with an eye on the long-term goal (Fay et al. 2015; Vogt-Schilb, Meunier, and Hallegatte 2018). Similarly, the sequence of the most feasible interventions is unlikely to deliver the change in political economy needed to make structural change possible. Instead, when choosing policies, countries should balance these policies' short-term political feasibility with their contribution to long-term objectives, including through a transformation of the political economy context.

How to do so? This chapter proposes an approach to help governments sequence policies that are feasible but that also build greater political support and reduce the costs of climate action over time, leveraging reinforcing policy feedback processes and targeting tipping points to accelerate transformational change toward net zero.

Because policy and political processes are not static, policy packages need to evolve over time in a dynamically efficient way. As well as being complex and multifaceted, the transition to net zero will occur over a significant time frame. When countries introduce new climate policies, they create effects that alter the broader political economy, influencing the types of policy they can introduce later. As such, it helps to think of the costs and benefits of climate policies in terms of dynamic efficiency—that is, the efficient allocation of resources over time. Whereas static efficiency focuses on how resources are used at a single point in time, a dynamically efficient appraisal considers how resource allocation today affects the availability and productivity of resources in the future. This is a common consideration for innovation and industrial policies that may deliver lower emission reductions per dollar invested than alternative options but that reduce green technology costs and make larger reductions possible and affordable in the future. This book extends that concept to the political economy: whereas the lowest-cost option today may lead to political backlash and create higher costs in the future, choosing a more expensive policy today might be more dynamically efficient if it shifts the political economy to make it easier to implement more efficient policies later.

This book therefore recommends appraising policies not only on their costs and feasibility but also on how they influence the political economy, build capacity to create new policy possibilities, and unleash the potential for transformative climate action. The first section considers policy choice from the perspective of institutional capacity. It explores how differences in capacity across countries can limit the ability to introduce different types of climate policy instruments and how, by making forward-looking policy choices, countries can build their institutional capacity to introduce more ambitious policies. The next sections consider policy selection through the lens of political feasibility and look at how countries can sequence policies to build greater political support for those that are initially less politically palatable. Although both dimensions (institutional capacity and political feasibility) are necessary for policies to be introduced, they require different strategies depending on countries' level of economic development and political context. Finally, and considering the need to catalyze rapid, transformational climate action, the chapter considers how governments can leverage reinforcing policy feedback processes and tipping points to build momentum and accelerate progress toward net zero.

Strategies to build greater institutional capacity to introduce climate policies

Over the last four decades, the number and variety of climate policies in different countries have steadily grown. According to the Climate Policy Database (box 3.1), countries have collectively introduced more than 4,500 climate policies since the 1980s. Categorizing these policies into over 50 different policy instrument categories provides a rich data set of policy pathways to learn from.

BOX 3.1

The Climate Policy Database

The Climate Policy Database,[a] maintained and frequently updated by the NewClimate Institute, provides one of the most comprehensive data sets of climate policies. Incorporating several other global policy databases, such as the Climate Change Laws of the World and the Organisation for Economic Co-operation and Development's policy instruments database, it also draws on country reports and specific policy documents. It provides a useful classification of climate policies by sector, policy instrument type, and implementation status.

Because the Climate Policy Database focuses primarily on national mitigation-related policies, it has limited coverage of policies relating to adaptation and resilience. It also has greater depth and quality of available information for larger emitters and countries that are required to provide detailed reporting on their policy implementation, such as Annex I countries to the United Nations Framework Convention on Climate Change and Organisation for Economic Co-operation and Development countries (Nascimento et al. 2021).

a. NewClimate Institute, Climate Policy Database, https://climatepolicydatabase.org/.

Although the number of climate policies introduced across countries does not necessarily reflect emissions reduction, studies have shown a significant association. For example, analyzing legislative data from 133 countries between 1999 and 2016, Eskander and Fankhauser (2020) find that each new climate law is associated with a short-term (within three years) reduction of 0.78 percent and a long-term (beyond three years) reduction of 1.79 percent in carbon dioxide (CO_2) emissions per unit of gross domestic product. Altogether, climate laws reduced annual global CO_2 emissions by 5.9 gigatons of CO_2 in 2016, surpassing the United States' CO_2 output for that year. Similarly, when controlling for historical emissions, income per capita, and countries' rule of law, Nascimento and Höhne (2023) demonstrate that having a higher number of climate policies is associated with lower projected emissions.

Different policies have had different impacts on emissions, even though it is not the only lens to evaluate them with. Figure 3.1 shows a variety of policy instruments and the associated effects on emissions across countries (controlling for per capita income, rule of law, industrial structure, and other important characteristics). Despite the difficulty in establishing causal relationships because of the simultaneous introduction of multiple policies and the complementarity across them, legally binding emissions reduction and renewable energy targets are associated with the largest reductions in emissions across countries. Introducing a binding greenhouse gas target is associated with a 10 percent reduction in annual CO_2 emissions per unit of gross domestic product in the short run (within three years) and a 22 percent reduction over the longer term; a carbon tax policy is associated with a 3.5 percent reduction in the short term and a 7.8 percent reduction over the longer term.

Even if they do not have a direct impact on emissions, policies that remove barriers to decarbonization or build essential capacity in the country can still be critically important climate policies. Although several policy instruments show much smaller effects on emission reductions, it does not mean they are useless or inefficient. Because different climate policy instruments act on various types of market failures, they can support the decarbonization process without necessarily reducing emissions directly. For example, strategic planning may not reduce emissions directly, but it can improve coordination

FIGURE 3.1. Estimated effects of different policy instruments on reducing CO_2 emissions

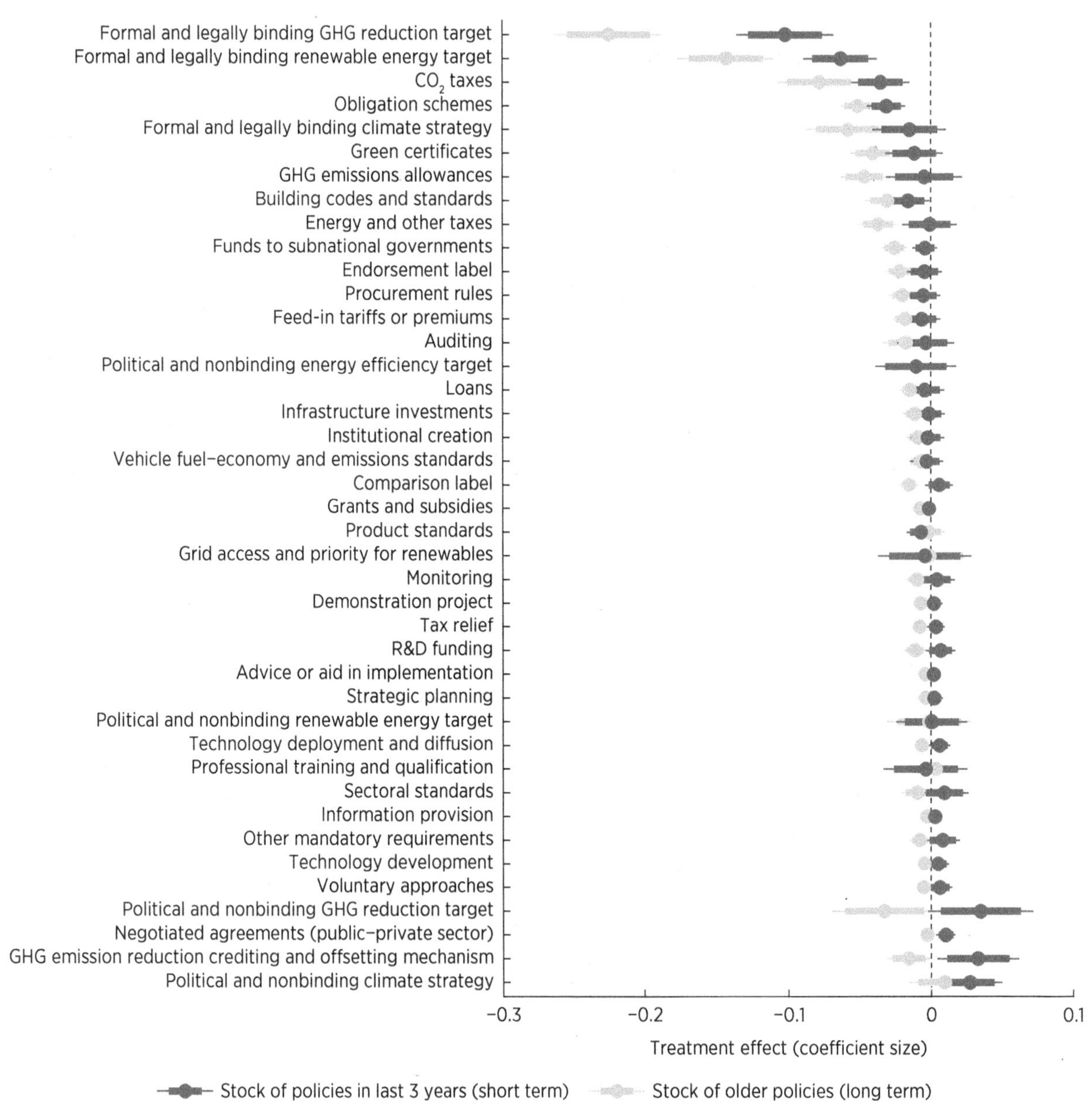

Source: Mealy et al., forthcoming.
Note: Two-way fixed effects and controls included. CO_2 = carbon dioxide; GDP = gross domestic product; GHG = greenhouse gas; $MtCO_2e$ = million tonnes of carbon dioxide equivalent; R&D = research and development.

across other instruments and increase their efficacy. Monitoring systems can play a similar role: because they improve policies over time and prevent inefficiencies, they can be necessary to a successful transition without reducing emissions directly. Other policies not evaluated here because they do not relate to climate change, such as reforms of financial or land markets or generic research and development support, can act in similar ways. And, as discussed later, some policies may be important to open the door to higher-efficacy policy—for example, nonbinding greenhouse gas emission targets can be an important step toward the introduction of binding targets.

Limitations in institutional capacity can restrict the number and breadth of climate policies that countries can introduce. Figure 3.2 shows various types of policy instruments adopted across countries represented as a heat map. Each column corresponds to a given country, and each row corresponds to a policy type. Blue squares signify that a country has announced a particular climate policy instrument, whereas white indicates that it has not. The figure has a characteristic triangular pattern, showing that some countries—such as Germany, India, Japan, and the Republic of Korea, shown in the far-left columns—have introduced many different types of policies, and that others, shown in the far-right columns, have introduced very few. More interestingly, policies that only a handful of countries have introduced (shown in the bottom rows) tend to appear in the columns of countries that have introduced a wide diversity of policy instruments. This suggests that it may be possible to introduce certain policies only in countries with sufficient administrative and policy-making capacity (Mealy et al., forthcoming).

Climate policy making is path-dependent, with institutional capacity limitations restricting the types of policy countries can introduce (Mealy et al. forthcoming). Policies are much easier to introduce if they build on prior related institutional capacity and know-how. For example, a country would have difficulty effectively implementing vehicle or industrial air pollution standards without first having the capabilities to monitor and audit vehicle or industrial performance. And countries with no prior experience of emissions monitoring or reporting—or with no form of market-based mechanism for reducing pollutants—may struggle to introduce carbon pricing. When considering the design of policy packages, governments should take such path dependency into account, because the choices they make today will influence their policy options tomorrow.

The Climate Policy Space provides a visual representation of such relationships, to better understand how policies and measures can build on each other. The Climate Policy Space (figure 3.3) is a network in which nodes represent climate policy instruments

FIGURE 3.2. **Triangular (nested) distribution of climate policies announced by countries**

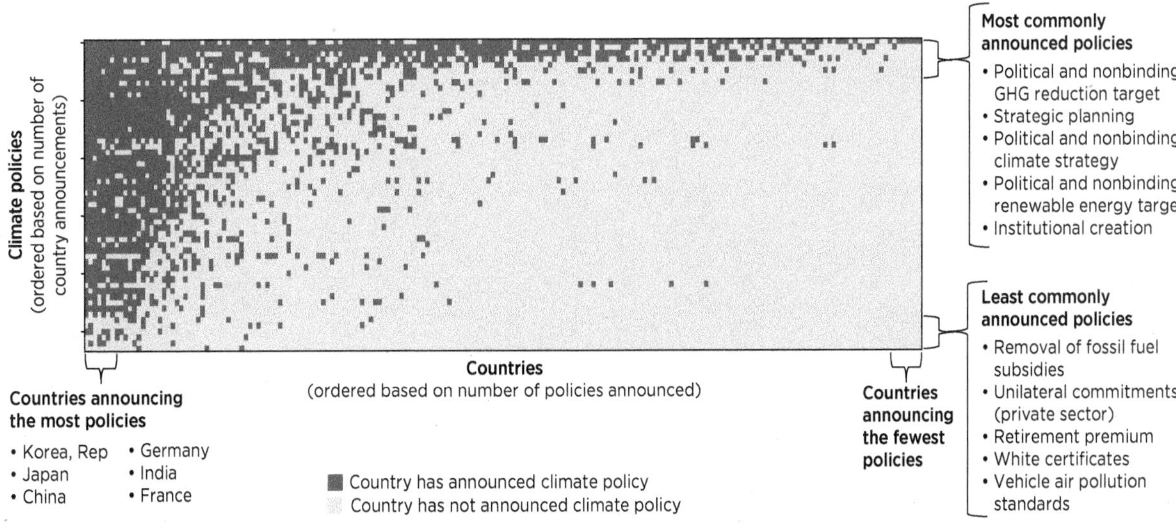

Source: Mealy et al., forthcoming.
Note: GHG = greenhouse gas.

FIGURE 3.3. The Climate Policy Space

a. The Climate Policy Space
Climate policy prevalence
High
Low
Removal of fossil fuel subsidies
User charges
Procurement rules
Public voluntary schemes
Unilateral commitments (private sector)
Technology deployment and diffusion
Demonstration project
Technology development
Advice or aid in inplementation
Funds to subnational governments
Information provision
Green certificates
Retirement premium
Vehicle air pollution standards
Sectoral standards
R&D funding
GHG emission reduction crediting and offset mechanism
Industrial air pollution standards
GHG emission allowances
CO_2 taxes
Monitoring
Grants and subsidies
Net metering
Auditing
Infrastructure investments
Loans
Coordinating body for climate strategy
Building codes and standards
Energy and other taxes
Feed-in tariffs
Tax relief
Grid access and priority for renewables
Obligation schemes
Institutional creation
Formal and legally binding GHG reduction target
Strategic planning
Formal and legally binding renewable energy target
Political and nonbinding renewable energy target
Political and nonbinding GHG reduction target
Political and nonbinding climate strategy
Formal and legally binding climate strategy
Formal and legally binding energy efficiency target
b. Key policy clusters
Mostly technology, innovation, and miscellaneous rare policies
Mostly regulatory, market-based instruments and enabling policies
Nonbinding targets and strategy
Binding targets and institutional creation
Nonbinding GHG reduction target
c. Climate policy prevalence for countries at different income levels
Climate policy prevalence
High
Low
Low-income countries
Lower-middle-income countries
Upper-middle-income countries
High-income countries

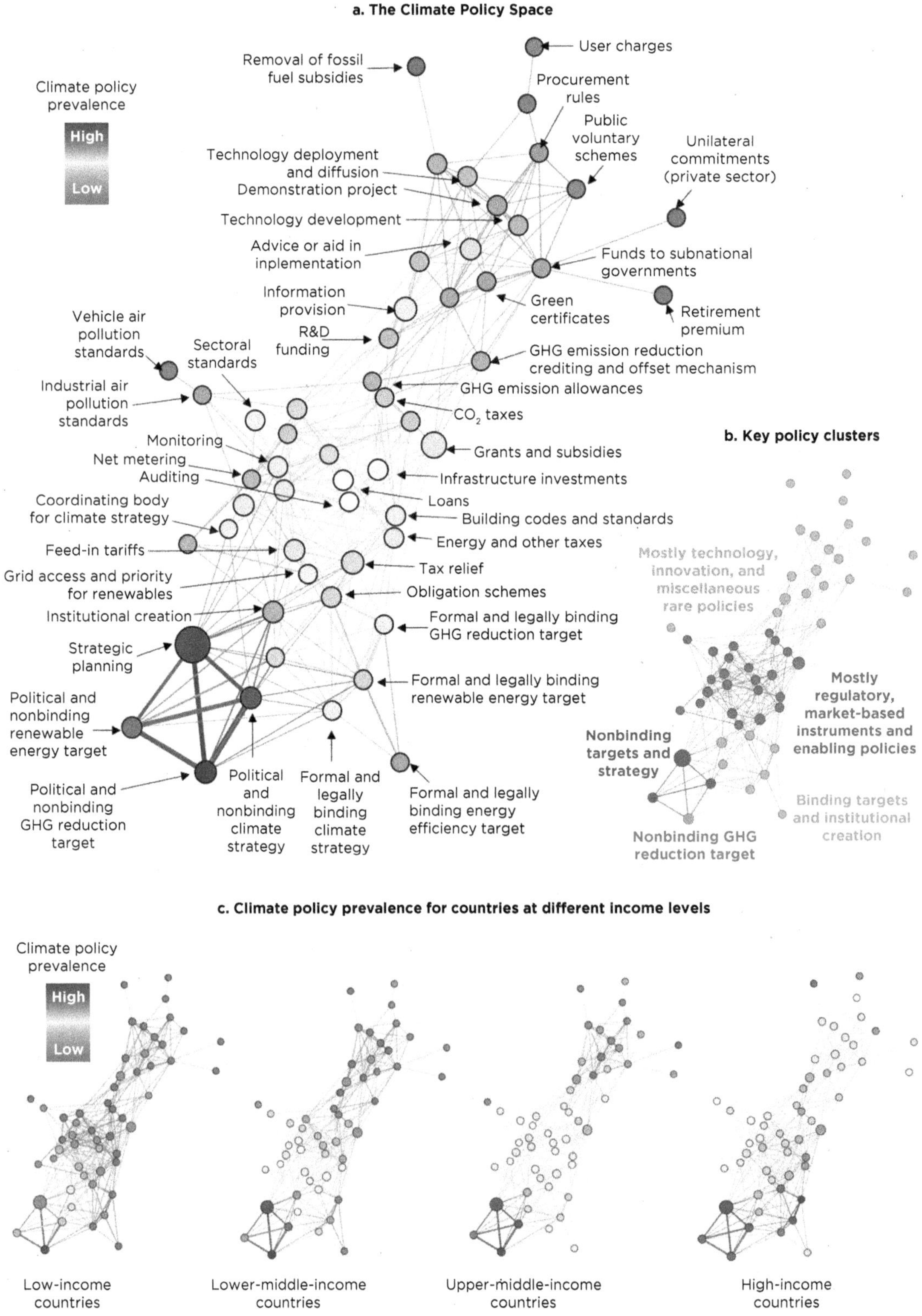

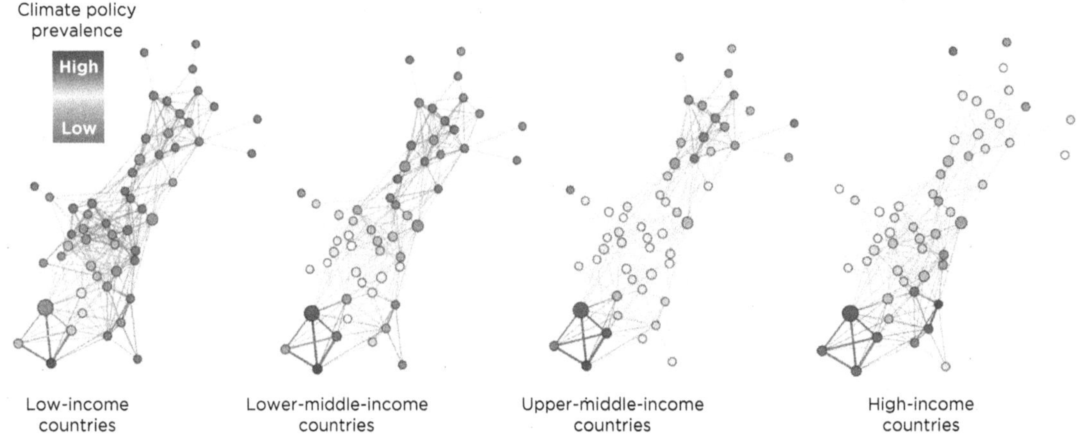

Source: Mealy et al., forthcoming.
Note: CO_2 = carbon dioxide; GHG = greenhouse gas; R&D = research and development.

linked to each other according to their relatedness in institutional capacity (see box 3.2 for more detail). In figure 3.3, panel a, climate policies are colored according to their prevalence across countries (purple policies have been more commonly introduced); in panel b, they are colored according to key policy clusters.

In panel b of figure 3.3, the pink and purple clusters at the bottom of the network consist of highly prevalent nonbinding targets and climate strategies that are fairly easy to introduce. These targets and strategies do not necessarily commit a country or constituents to do anything. The turquoise cluster comprises binding targets and the creation and strengthening of required climate institutions, representing a natural step up from nonbinding targets in a country's policy pathway. The blue cluster in the middle largely consists of regulatory instruments, such as product and industry standards; market-based instruments like carbon, energy, and other taxes; and key enabling policies related to auditing, monitoring, and coordinating bodies for climate strategy. Finally, the olive cluster includes a variety of policy instruments relating to technological deployment and innovation and other rare policies that tend to be found only in countries that have previously introduced a wide diversity of policy instruments.

Countries with different income levels are concentrated in different regions of the Climate Policy Space. Panel c of figure 3.3 shows that low-income countries, which typically have less developed levels of institutional capacity, tend to have introduced climate policies found in the lowest clusters of the network. Lower-middle- and upper-middle-income countries show a broader range of policies in the turquoise and blue clusters, suggesting that expanding policy-making capacity into binding targets, institutional creation, and regulatory and market-based instruments may go hand in hand with rising levels of economic development. High-income countries span a vast range in the Climate Policy Space network, with a notable presence in the olive cluster that focuses on unique and technology-centric policies, arguably the actions that require the most capacity.

Countries move through the Climate Policy Space in predictable ways because past climate policies influence future policies. They show a tendency to introduce new

BOX 3.2

Calculating policy relatedness

Countries will likely find it easier to introduce a policy if they have prior experience introducing policies that involve similar (or related) institutional and administrative capacities and requirements. Unfortunately, data constraints make it challenging to directly measure specific institutional capacity requirements consistently across countries.

Mealy et al. (forthcoming) propose a novel approach for estimating the relatedness of institutional capacity between two climate policies by exploiting the pattern of climate policy co-occurrence within countries. Specifically, two policies are assumed to require similar underlying institutional capacity and know-how if they are more likely to co-occur within countries. The measure is also weighted, with countries that have introduced many policies assigned a lower weight in the overall calculation. Previous studies have applied such techniques to analyze the path dependence of economic development across countries and regions (Hidalgo et al. 2007; Mealy and Coyle 2022; Zaccaria et al. 2014).

Mealy et al. (forthcoming) show that the more related a new climate policy is to a country's existing set of policies, the more likely that country will be to introduce this policy in the next five years, even after accounting for important country characteristics such as income levels, rule of law, government effectiveness, and carbon dioxide emissions.

policies that are connected to existing policies in the Climate Policy Space network, building on prior related know-how and institutional capacity. Mealy et al. (forthcoming) show that this tendency is statistically significant across countries, even after controlling for important country characteristics such as income levels, population, corruption, rule of law, and government effectiveness, and can be used to make predictions about the types of climate policy that countries are more likely to introduce in the future. As such, the set of climate policies implemented at one time affects in a measurable way the probability of implementing further policies in the future.

Climate Policy Feasibility Frontiers (CPFFs) can help inform policy choices that realistically work with countries' current policy-making capacity and successively build greater capacity to introduce more ambitious types of policy. Combined with the usual analysis of the efficacy, costs, and benefits of policies, the CPFF can help identify the most promising policies in a given country context. It consists of two key dimensions:

1. *Relative likelihood of introducing a policy in the next five years.* Based on how related a new policy is to a country's existing policies, this metric is expressed in relative terms, comparing policies without measuring their absolute likelihood. This metric measures the ease of implementing a given policy, based on a country's prior policy experience and inferred policy-making capacities.
2. *Capacity-building potential* aims to capture the learning and capacity development potential associated with the introduction of a new policy. This metric measures how the introduction of a given policy is expected to change a country's institutional capacities, making it easier to implement other climate policies in the future.

Figure 3.4 shows the current positions in the Climate Policy Space and CPFFs for Türkiye and Viet Nam. Each dot in the CPFF denotes a new policy not introduced before and corresponds to policies colored in gray in the Climate Policy Space. The figure also shows the emissions reduction potential of these policies, based on the analysis of past policy adoptions and associated changes in emissions presented in figure 3.1. With the costs and benefits, the efficacy of these policies is a crucial dimension to consider when prioritizing and designing policies.

The CPFF identifies policies that may be easier to implement, as illustrated by a set of policies shown in dark bold circles in panels a (for Türkiye) and c (for Viet Nam). For Türkiye, for example, policies that appear the easiest to introduce in the next five years include a legally binding climate strategy or a binding emissions reduction target, which would require including such a strategy or target in a law, such as a climate change framework law.[1] Türkiye's CPFF shows that a legally binding climate strategy is about 30 percent more likely to be introduced in the next five years compared to the introduction of a retirement premium, which is far away from Türkiye's existing policies in the Climate Policy Space.

The CPFF can also help in the design of a step-by-step policy pathway, going beyond short-term opportunities and ensuring progress toward long-term goals. For example, the CPFF identifies an emissions trading scheme (ETS) or carbon tax as potential next steps for Türkiye. These two policies are close to each other in the policy network, showing they are often introduced together and that the two instruments more often complement each other than substitute each other. In practice, introducing either or both instruments depends on the political context. Partly because of Türkiye's geographic proximity with the European Union and its ETS, the government of Türkiye has recently

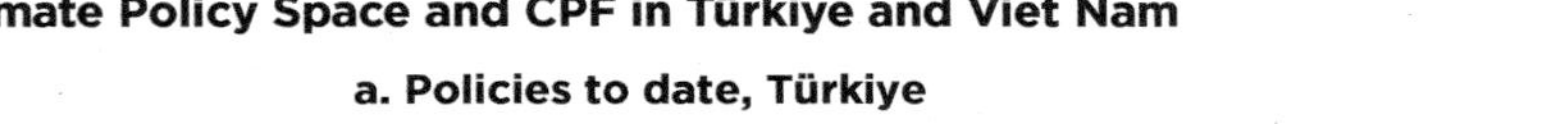

FIGURE 3.4. Climate Policy Space and CPF in Türkiye and Viet Nam

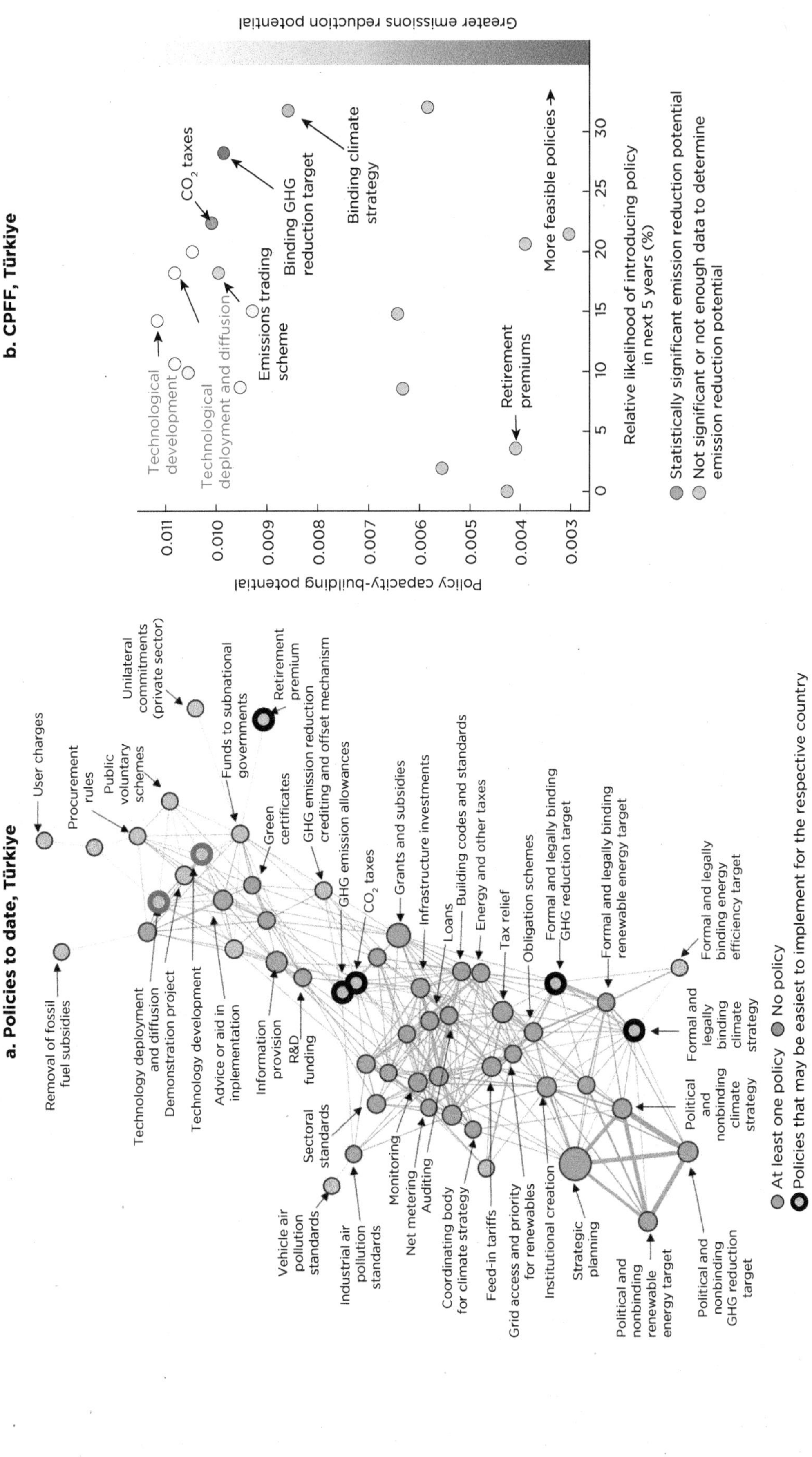

(Continued)

FIGURE 3.4. **Climate Policy Space and CPFFs, Türkiye and Viet Nam** (continued)

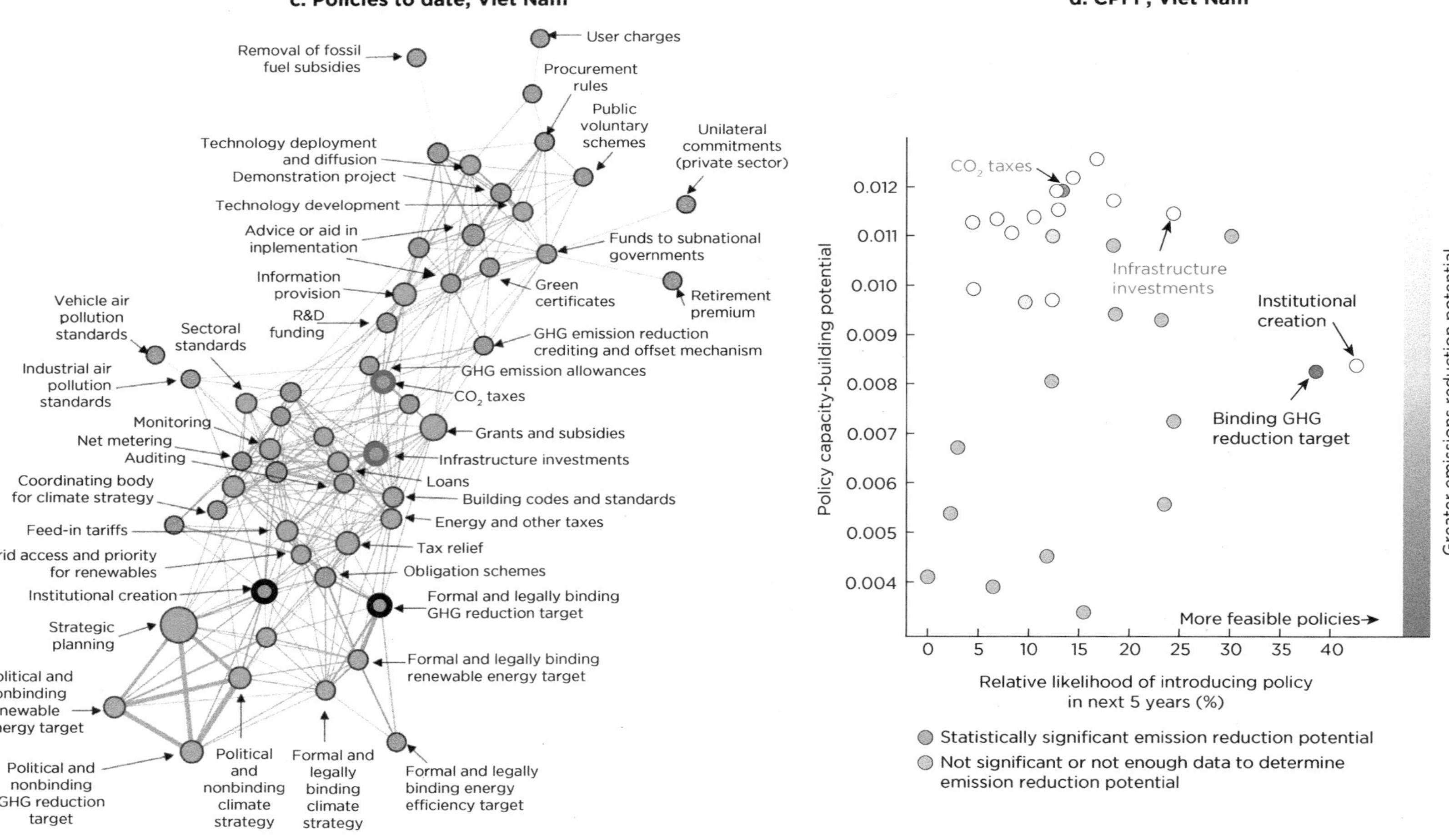

Source: Mealy et al., forthcoming.
Note: CO_2 = carbon dioxide; CPFF = Climate Policy Feasibility Frontier; GHG = greenhouse gas; R&D = research and development.

announced the future implementation of an ETS, confirming the CPFF assessment that an ETS is a highly feasible intervention in the country.

The CPFF also identifies technological deployment and diffusion and technological development as "stretch" policies for Türkiye. Although these policies (labeled in purple in the CPFF and highlighted in a purple circle in the Climate Policy Space) are less aligned with Türkiye's existing policies and potentially more difficult to introduce, they have the potential to build the country's capacity to deploy low-carbon technologies and engage in green innovation and technological development. In addition to their climate benefits, such capacities could also have important economic advantages and build momentum toward more rapid climate action.

The CPFF emphasizes the importance of the country context and identifies different recommendations across countries. Viet Nam has a different set of institutional capabilities from that in Türkiye. For Viet Nam, the most feasible and likely policies include institutional creation and binding greenhouse gas targets. Policies to boost climate-related infrastructure investments could help Viet Nam build further capacity. Moreover, although carbon pricing policies appear to be more challenging in the short term, other policies can help pave the way toward this goal.

Sequencing to build greater political support for climate action

Policies that can deliver immediate benefits to key groups and the economy more broadly can be politically easier to implement and can help build support for further climate action. With the public usually favoring "pull" over "push" policies (Drews and van den Bergh 2016), tax relief, grants, and subsidies tend to be among the first policies implemented in the upper-middle-income and higher-income countries that can afford them (Meckling, Sterner, and Wagner 2017). When these policies foster supportive coalitions and broader public support—as is the case with renewable energy support policies—they can build momentum for more ambitious and less politically palatable policies down the road (Pahle et al. 2018)—see box 3.3.

Strategically sequencing policies can grow political support over time and shape a political economy that is more conducive to climate action. For example, in Germany, early renewable energy research and development funding, subsidies, and capacity targets created the basis for low-carbon energy interests to emerge, drove down technology costs, created synergies with other energy and development goals, and increased political and public support for low-carbon power—despite the challenges of establishing renewable energy businesses in the country. These actions, in turn, enabled more ambitious energy decarbonization policies. The case of Viet Nam, which mobilized private financing for 20 gigawatts of renewable energy in record time, shows how feed-in tariffs and other "pull" instruments can deliver impressive results in the short run (Do et al. 2021). But such instruments need to evolve to become financially sustainable. In Viet Nam, financial losses in 2022 and 2023 suggest a lack of financial sustainability and reduce attractiveness for private capital. Countries may therefore need complementary measures to promote transparency in transactions, more sustainable terms, and the ability to crowd in cutting-edge technology. However, policies that distribute benefits might still encounter opposition, especially when powerful vested interests understand the long-term impacts that the subsidies could have on them.

BOX 3.3

China's sustainable energy transition policy sequence

In their study of the evolution of China's national sustainable energy policy mix over 40 years, Li and Taeihagh (2020) show how sequencing helped reduce resistance from existing institutions and increase support by fostering winning coalitions, while gradually increasing policy stringency and reducing costs (figure B3.3.1).

First, the government provided 10 years of support for wind and solar photovoltaic, building interest groups in low-carbon technologies. It then gradually ratcheted up policy stringency by increasing emissions charge rates and tightening emissions limits and air quality standards, eroding the incumbency of coal power. Finally, it started reducing feed-in tariffs and other government subsidies after supporting grid-parity renewable energy. Together, these actions reduced the cost of China's climate policies.

FIGURE B3.3.1. **Evolution of China's environmental policy mix, 1980–2020**

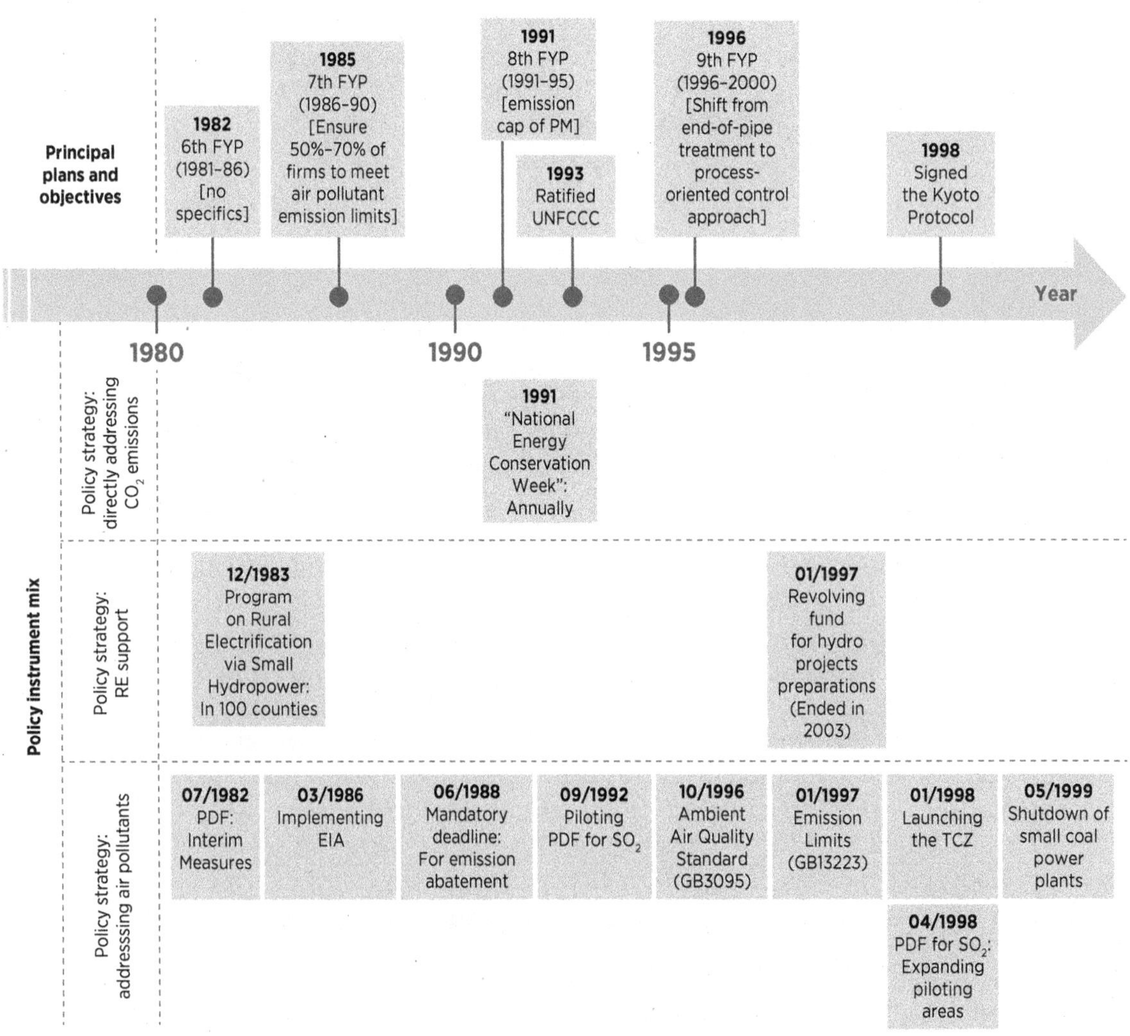

(Continued)

BOX 3.3

China's sustainable energy transition policy sequence (continued)

FIGURE B3.3.1. Evolution of China's environmental policy mix, 1980–2020 (continued)

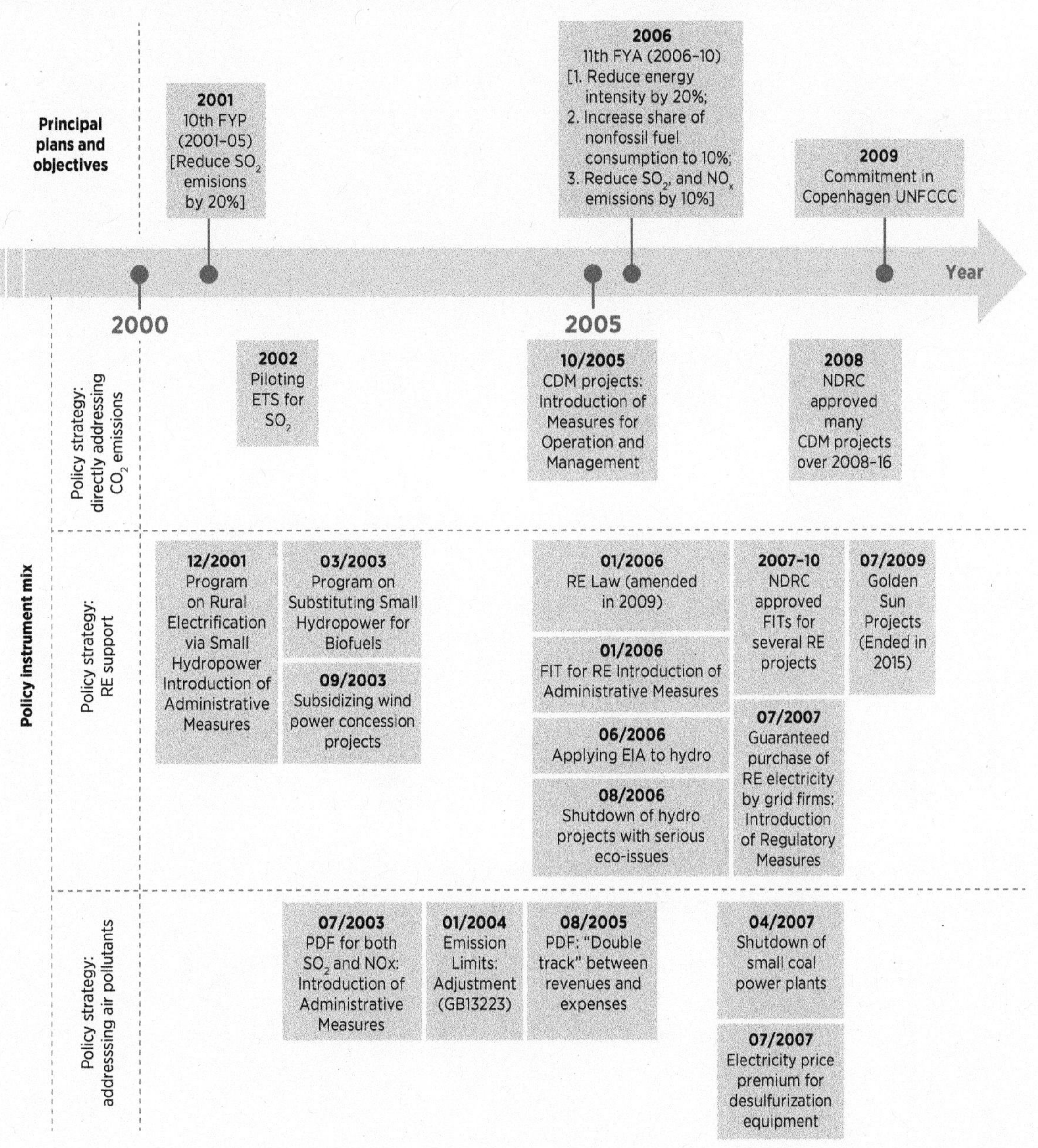

(Continued)

BOX 3.3

China's sustainable energy transition policy sequence (continued)

FIGURE B3.3.1. **Evolution of China's environmental policy mix, 1980–2020** (continued)

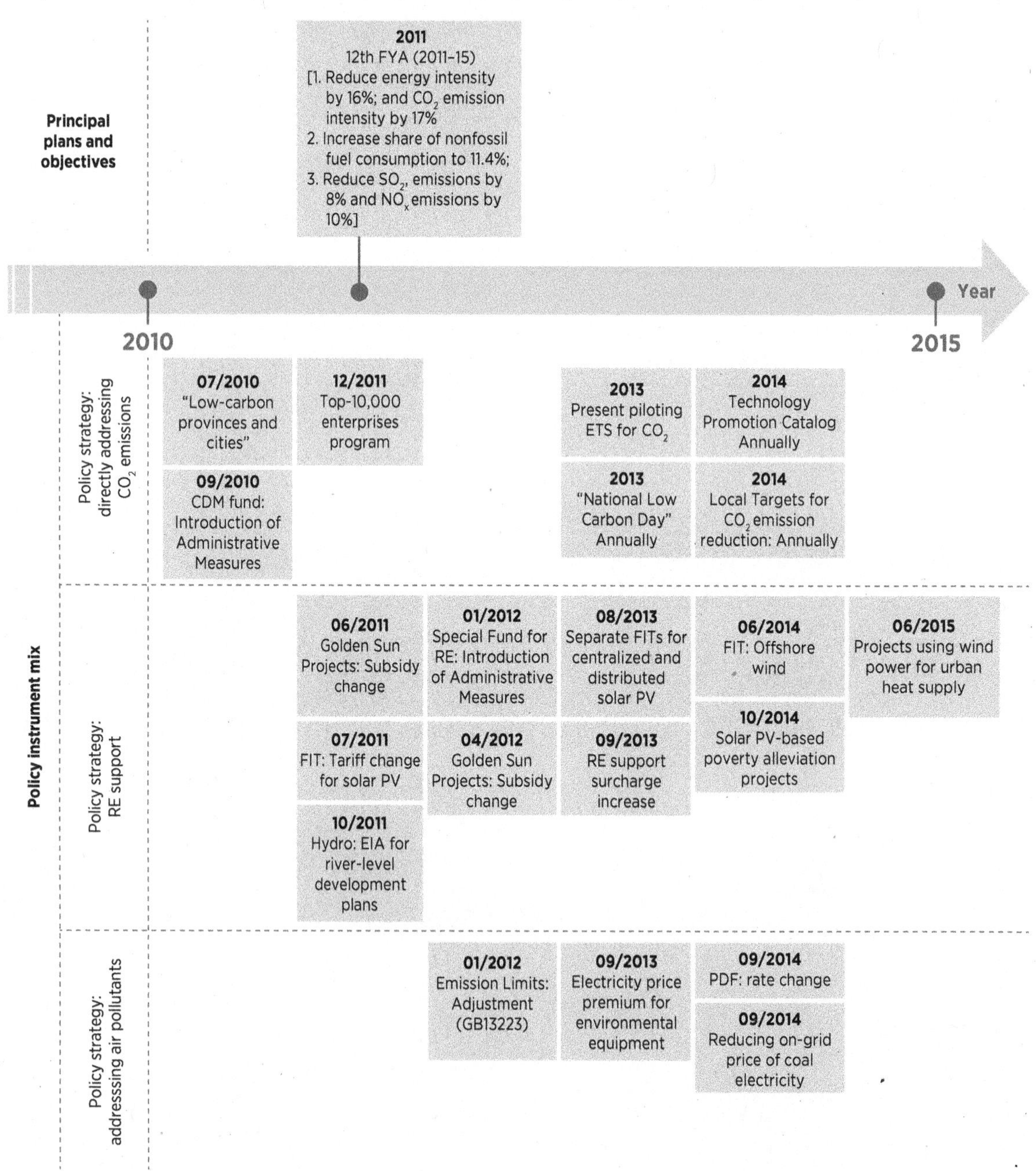

(Continued)

BOX 3.3

China's sustainable energy transition policy sequence (continued)

FIGURE B3.3.1. **Evolution of China's environmental policy mix, 1980–2020** (continued)

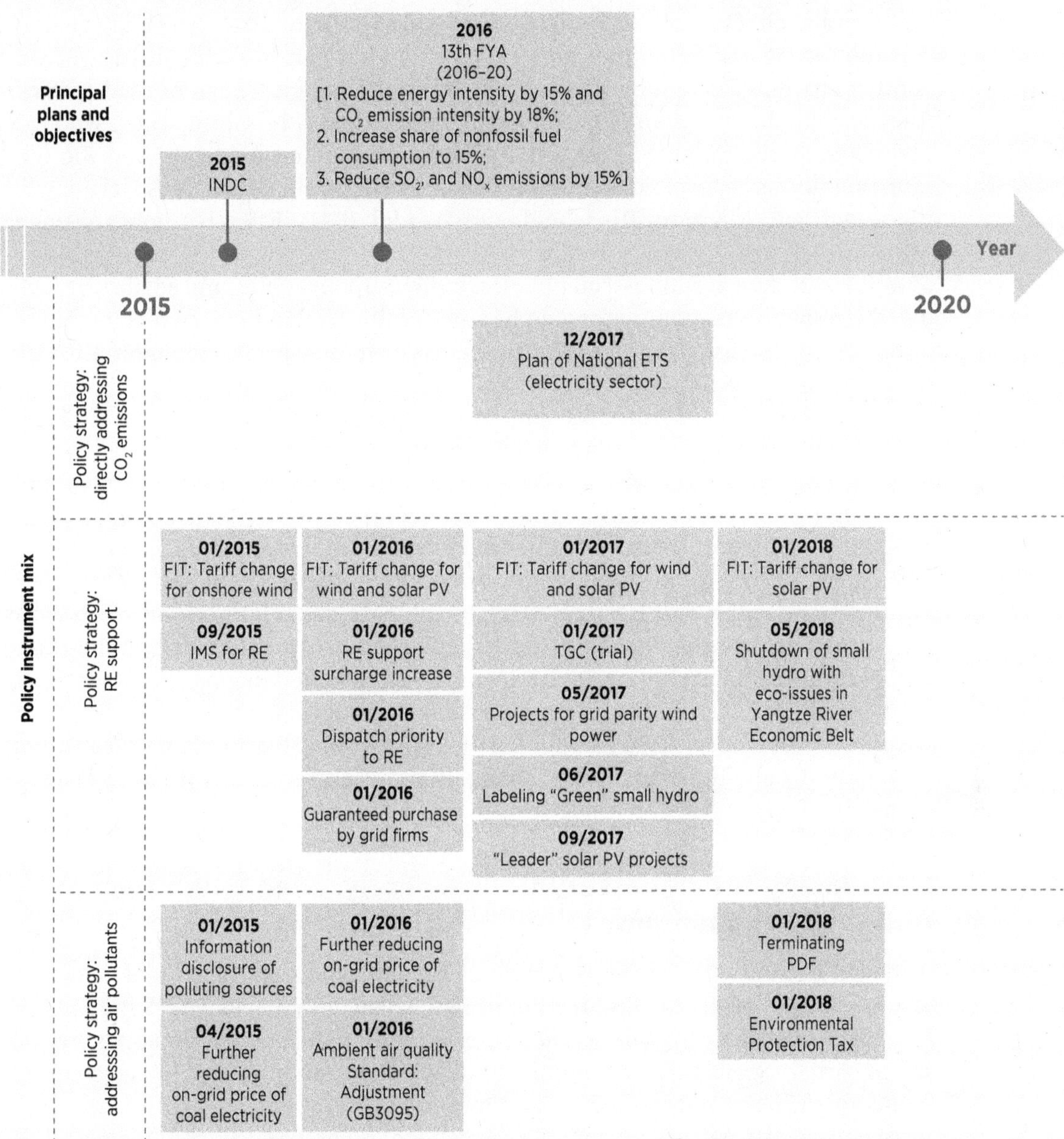

Source: Based on Li and Taeihagh 2020.

Note: China's national ETS became operational in 2021 and focuses on the regulation of power sector companies. It uses allowances freely allocated and based on benchmarks considering actual production levels. The system is constantly being further developed, and its coverage is planned to expand to other sectors as well. With more than 4 billion tonnes of CO_2 covered—accounting for over 40 percent of the country's carbon emissions—the ETS is the world's largest in terms of covered emissions (ICAP 2023). CDM = clean development mechanism; CO_2 = carbon dioxide; EIA = environmental impact assessment; ETS = emissions trading system; FIT = feed-in tariff; FYP = five-year plan; GB = *Guobiao*, Chinese for "national standard"; IMF = Information Management System; INDC = intended nationally determined contributions; NDRC = National Development and Reform Commission; NOx = nitrogen oxide; PDF = pollutant discharge free; PM = particulate matter; PV = photovoltaic; RE = renewable energy; SO_2 = sulfur dioxide; TCZ = two control zones; TGC = Tradable Green Certificate; UNFCCC = United Nations Framework Convention on Climate Change.

Green industrial policies offer a good starting point in a policy sequence, because they can foster winning coalitions that provide a useful political support base. Many have advocated for starting with green industrial policies, such as research and development support or subsidies, and green trade policies, such as support for low-carbon export industries, to create coalitions of actors with economic interests in low-carbon industry and climate policy action (Cullenward and Victor 2020; Meckling et al. 2015). The unprecedented fall in the cost of solar and wind—unthinkable just 10 years ago—is often attributed to a mix of industrial and trade policies, such as feed-in tariffs in Europe and direct credit support for manufacturers in China (Lockwood 2022; Meckling, Sterner, and Wagner 2017; Pahle et al. 2018). This has paved the way for more ambitious renewable energy targets and other policies, such as Germany's coal phase-out law (Markard, Rinscheid, and Widdel 2021). Effective policy sequencing can also weaken the power of potential climate policy opponents (Nacke, Cherp, and Jewell 2022). Sunrise industrial policies, which support emerging sectors or technologies, are sometimes complemented by sunset industrial policies, which facilitate and organize the downscaling of declining sectors or industries (see chapter 4 for a more in-depth discussion).

Taking advantage of windows of opportunity can make it easier to introduce politically challenging policies. The sustainability transitions literature highlights windows of opportunity as critical points when conditions become more favorable for changing incumbent or locked-in institutional landscapes (Geels 2006, 2012; Mealy et al. 2023). Different factors can open such windows—for example, it can be easier to implement fossil fuel subsidies, which are politically difficult, when global oil prices are low (Rentschler and Bazilian 2017), or climate fiscal policies or green recovery packages when a crisis response suddenly increases thresholds for public spending and government intervention in the economy, as observed during the COVID-19 pandemic (Dafnomilis et al. 2022). Highly visible natural disasters such as the 2003 heat wave in Europe or the 2005 landfall of Hurricane Katrina in New Orleans, both at least partially attributed to climate change, have also affected the political economy of climate policies.

Introducing the necessary policies early, even when they are politically challenging

Although starting with policies that are easier to implement can help build momentum, it is sometimes necessary to introduce politically challenging policies, and doing so earlier on in a policy sequence can be desirable. For example, introducing a carbon price, even if it is low or partial, signals a clear commitment, allows actors to adjust—say, by switching investments to efficiency improvements and lower-carbon equipment—and sets the basis for price or coverage increases as actors' expectations and preferences shift (Sato et al. 2022). Some administratively demanding policies, usually less likely to be applied in the early stages, could also be worth implementing early to help align actors' expectations and build institutional capacity. For example, clear efficiency or emissions labeling of appliances, vehicles, and other products and services creates awareness among citizens and may help shift the range of policies that are acceptable to the general population at a given time (Kelsall et al. 2022; Rosenbloom, Meadowcroft, and Cashore 2019).

Preventing high-carbon path dependence is often a good investment, especially in low-income environments, even if doing so is difficult or expensive. Always taking the easy option could decrease the feasibility of more ambitious policies in the decades to come, leading countries to carbon-intensive physical or institutional lock-in. This is a major risk in countries still building a lot of infrastructure, and policy decisions could put them on a low-carbon development trajectory or lead them to carbon lock-in. In sectors that are particularly expensive and difficult to decarbonize, like transportation, starting early will make the transformation as progressive and smooth as possible, minimizing long-term costs. Starting with the most expensive option today sometimes makes sense in the long term (Vogt-Schilb, Meunier, and Hallegatte 2018).

Urbanization patterns represent one of the largest sources of lock-in, justifying early action. Even though low-income cities have relatively low emissions, if they grow with low density and a high reliance on individual vehicles, they will struggle to develop and implement the transportation decarbonization policies they will need in the future. They will also struggle to ensure efficient and attractive public transportation, which is more important at higher income levels. And, despite the difficulty in implementing transit-oriented, low-carbon development policies today, doing so could make it much easier to decarbonize transportation in the coming decades without major trade-offs for development (Avner, Rentschler, and Hallegatte 2014). If people enjoy good levels of mobility and accessibility in urban spaces with public or active transportation, they are less likely to oppose policies that increase the cost of car use or phase out fossil fuel vehicles—because they have readily available alternatives. Support for, or acceptance of, policies that limit car use is higher in European and Scandinavian cities with well-developed public transportation infrastructure (Kuss and Nicholas 2022; Mareschi et al. 2022).

Building policy ambition and stringency through feedback and tipping points

Strategically selecting and sequencing feasible policies to build greater institutional capacity and political support does not mean climate progress will be slow. By taking advantage of the dynamism of socio-technical-political systems, governments can build momentum to accelerate transformational climate action. Each step a government takes and each policy it implements can reinforce further climate action in the future.

Introducing specific policies can transform the associated politics, which in turn shapes the future space of policy possibilities. *Policy feedback* relates to the effects that policies can have in either reinforcing or undermining the direction or pace of future policy making. The adage "new policy creates new politics" captures the way that each climate policy or intervention affects the political economy landscape, creating new incentives, spreading new ideas, supporting new coalitions, and reforming institutions.

Some policies can drive positive, reinforcing feedback effects, leading to faster climate progress and more ambitious action. Climate policy feedback effects are a key theme in the sociotechnical system transition literature, with special attention paid to links and interdependencies between low-carbon technology innovation and scaling, on the one hand, and changes in social and political values, norms, discourses, and behaviors, on the other (Aklin and Urpelainen 2013; Geels et al. 2017; Lockwood 2015). Green industrial policy is essentially a feedback-based strategy that directs benefits or "rents" toward green industries, growing them and using them to increase political support more broadly (Meckling et al. 2015). Such policies can trigger positive policy feedback effects that drive institutional processes toward deeper or faster green reforms. In so doing, they can increase the base of support for climate policy over time as more winners emerge, from green industry shareholders and investors to their employees and labor representatives, the communities where they are based, their local political representatives, the customers who use their products, and so on.

Factoring in policy feedback effects is an essential complement to cost-benefit assessments of economic efficiency and other approaches that countries use to inform their policy choices. Putting a political economy lens on policy sequencing shows that, although second- (and even third- or fourth-) best policies based on static economic efficiency might cost more, especially at earlier stages, they can pave the way for more stringent and efficient policies later, improving the overall feasibility and cost of transformative policies over time (Pahle et al. 2018). From a dynamic efficiency perspective, it is as much about reducing barriers as strengthening enabling factors to ensure the political economy can shift from being a brake on transformative climate action to driving it. This is part of the logic that underlies the call for green trade and industrial policy, whereby direct support for green industry can generate increasingly powerful winning coalitions that champion more ambitious decarbonization policy (Meckling et al. 2015). Many countries and subnational areas—including early movers Germany and the US state of California—deployed targeted subsidies and other policy instruments to support renewable energy with great success (Pahle et al. 2018).

Governments can also aim to strategically target tipping points in social, technological, and political domains, which can drive rapid and systemic change. Climate policies that overcome inertia, reduce opposition, and increase support help to develop momentum behind a clear direction of change. In the power sector, for example, renewable energy combined with storage is expected to displace fossil fuels in the next decade even without additional policy support (IEA 2023). But going in the right direction does not guarantee that we will reach our desired destination—in this case, net zero—in time. Indeed, various assessments suggest that we are moving too slowly to meet climate goals (Boehm et al. 2022). Tipping points that can accelerate change are a critical component of a climate sequencing strategy (figure 3.5).

FIGURE 3.5. **Enabling conditions to trigger positive tipping points**

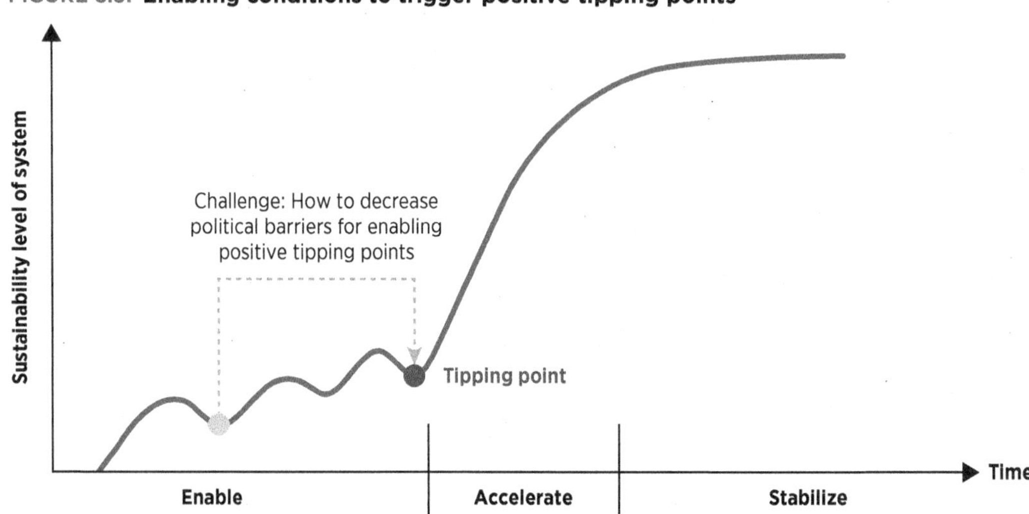

Source: Fesenfeld et al. 2022.

A *tipping point* refers to nonlinear change in a complex system, in terms of the speed or nature of change (Milkoreit 2022). The primary driver of a tipping point is the dominance of positive over negative feedback effects, which reinforces change. Once a tipping point has been breached, the likelihood of quickly or easily returning to a previous state or pace of change is low (Milkoreit 2022). Several types of tipping point are relevant for climate governance, transition strategies, and policy sequencing decisions.

- *Climate tipping points: major, rapid, and abrupt changes in the climate system.* Breaching climate tipping points would likely trigger further tipping points in ecological and human systems. Examples include major sea level rise resulting from collapsing ice sheets or dieback of important biodiverse biomes, such as the Amazon Rainforest (Armstrong McKay et al. 2022). Warnings from the scientific community that we are fast approaching climate tipping points are becoming more severe (OECD 2023). If those tipping points are triggered, it is reasonable to assume that the political economy of climate policy would change rapidly, perhaps leading to radically different approaches to managing climate action.
- *Social tipping points: rapid self-reinforcing shifts in attitudes, beliefs, behavior, and values in society.* Whether social tipping points are triggered depends to an extent on where change happens in a social network. People are more likely to respond to changes in the attitudes or behavior of influential actors—for example, a thought or political leader—because they are well-connected and more visible; but this response will be mediated by factors such as trust, credibility, and social identity (Sönke et al. 2022). The response can also be a question of critical mass. One experimental study suggests that, when just 25 percent of a population group changes its norms or behaviors, the majority can be "tipped" to follow (Centola et al. 2018). Another suggests, however, that the threshold is likely to differ across societies, depending on factors such as risk tolerance, nonconformity, and available incentives for early movers (Andreoni, Nikiforakis, and Siegenthaler 2021).
- *Technological tipping points: significant shifts in technology maturity, performance, costs, or accessibility.* Several key technological tipping points have already been crossed—for example, when renewable energy became the cheapest option to generate electricity in most markets.
- *Policy tipping points: rapid shifts in support for and implementation of a particular policy or set of policies.* Major crisis, scientific discovery or breakthrough, changes in public opinions and values, a policy paradigm shift, or a change in the balance of power between oppositional and supportive coalitions can all contribute to triggering a policy tipping point. Once this tipping point is breached, reinforcing feedback effects can accelerate diffusion—for example, across levels of government—and embedding, such as in social discourse. In the wake of a policy tipping point, an unpopular, controversial, or unfeasible policy or set of policies can become widely accepted and supported, making continued progress likely and reversal unlikely.

Sharpe and Lenton (2021) make the case that policies can accelerate progress by targeting "upward-scaling tipping cascades"—that is, progressive activation of tipping points that increase the likelihood of triggering another at a larger scale. This would create a sort of tipping point path dependence (figure 3.6). Although their study focuses on how small groups of countries working together can activate tipping points in the global economy, countries can apply findings to their national context. Looking at the power sector and road transportation vehicles, the authors zoom in on relative technology cost

FIGURE 3.6. Upward-scaling tipping cascades to meet climate goals

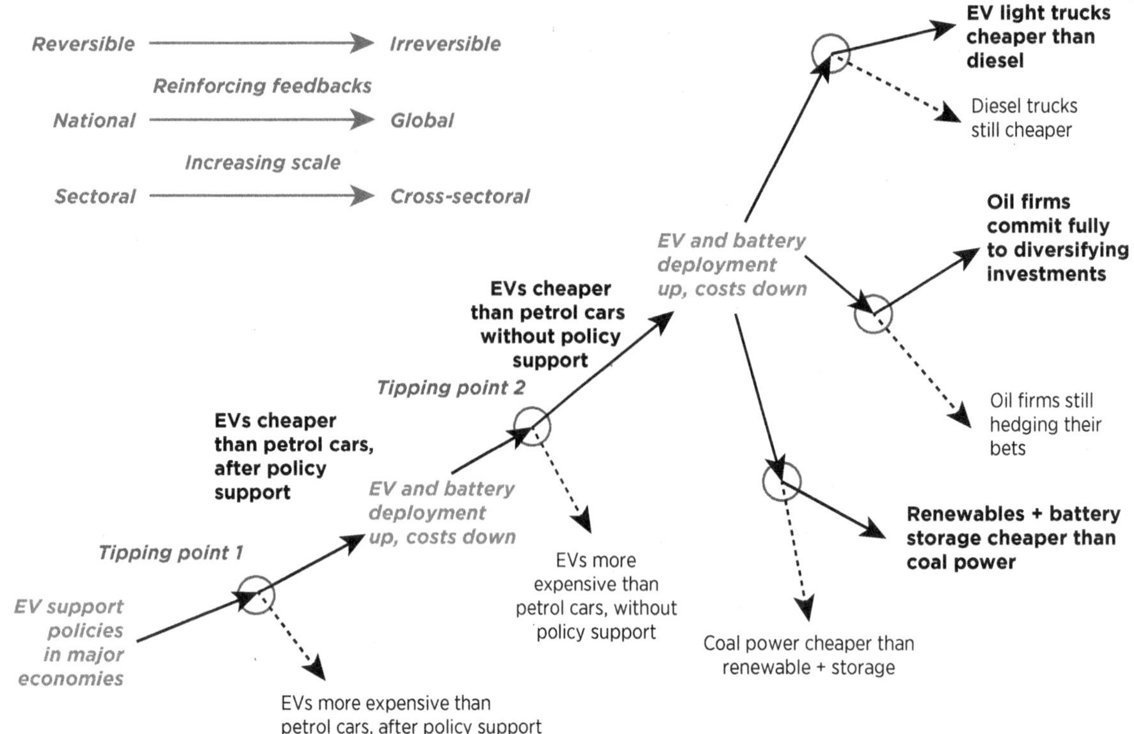

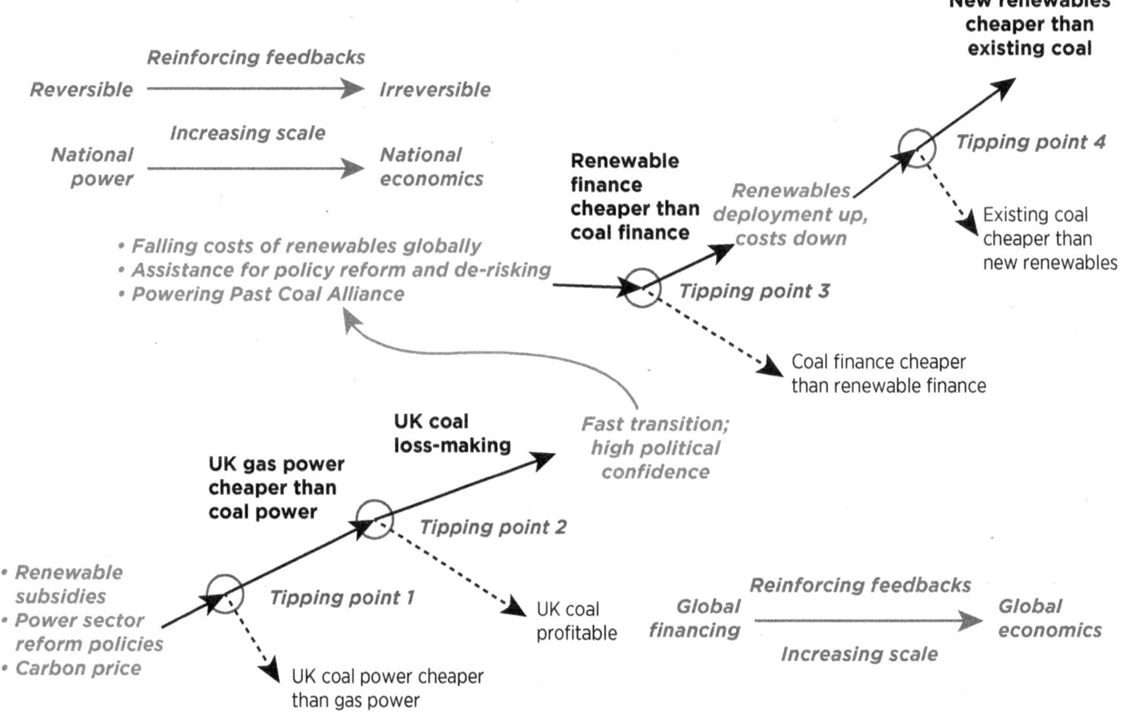

Source: Sharpe and Lenton 2021.
Note: EV = electric vehicle.

tipping points, when low-carbon technology (gas and then renewable energy in the power sector; electric vehicles in the transportation sector) become cheaper than high-emitting technology (coal and then gas in the power sector; internal combustion engines in the transportation sector), first with—and then without—policy support. They show how these tipping points lead to changes not only in technology shares but also in climate politics, enabling more ambitious and faster policy change.

Radical climate policies can help trigger positive tipping points. Policy (feedback) effects can contribute to path dependence, acting as a funnel for future policy options. Strategic policy sequencing can harness these effects to narrow the range of policy options to a more ambitious or more stringent set of options, including by fostering more policy winners. But a tipping point still needs to be triggered to accelerate change, and that triggering may require intentional forcing (Abson et al. 2017; Fesenfeld et al. 2022; van der Ploeg and Venables 2022). Van der Ploeg and Venables (2022) make the case for radical climate policies that can provide the necessary "big push," arguing that marginal policies, such as setting the price of carbon to the social cost of carbon, are unlikely to work as a trigger. Instead, they suggest that countries will need "big push" policies—for example, a sizable electric vehicle subsidy—to trigger tipping points, such as sociotechnical reinforcing effects that lead to a rapid electric vehicle diffusion and displacement of internal combustion engines. Although expected to be costly, such "big push" policies can be temporary, because they will no longer be necessary once the economy shifts toward the new, superior, equilibrium.

Note

1. Targets or strategies are considered "binding" when individuals and institutions in the public and private sectors must comply with them—for instance, because they are part of national legislation.

References

Abson, D., J. Fischer, J. Leventon, J. Newig, T. Schomerus, U. Vilsmaier, H. von Wehrden, P. Abernethy, C. D. Ives, N. W. Jager, and D. J. Lang. 2017. "Leverage Points for Sustainability Transformation." *Ambio* 46 (1): 30–39. https://doi.org/10.1007/s13280-016-0800-y.

Aklin, M., and J. Urpelainen. 2013. *Renewables: The Politics of a Global Energy Transition*. Cambridge, MA: MIT Press.

Andreoni, J., N. Nikiforakis, and S. Siegenthaler. 2021. "Predicting Social Tipping and Norm Change in Controlled Experiments." *Proceedings of the National Academy of Sciences* 118 (16): e2014893118. https://doi.org/10.1073/pnas.2014893118.

Armstrong McKay, D. I., A. Staal, J. F. Abrams, R. Winkelmann, B. Sakschewski, S. Loriani, I. Fetzer, S. E. Cornell, J. Rockström, and T. M. Lenton. 2022. "Exceeding 1.5°C Global Warming Could Trigger Multiple Climate Tipping Points." *Science* 377 (6611): eabn7950. https://doi.org/10.1126/science.abn7950.

Avner, P., J. E. Rentschler, and S. Hallegatte. 2014. "Carbon Price Efficiency: Lock-In and Path Dependence in Urban Forms and Transport Infrastructure." Policy Research Working Paper 6941, World Bank, Washington, DC. https://openknowledge.worldbank.org/handle/10986/18829.

Boehm, S., L. Jeffery, K. Levin, J. Hecke, C. Schumer, C. Fyson, A. Majid, and I. Jaeger. 2022. *State of Climate Action 2022*. World Resources Institute. https://www.wri.org/research/state-climate-action-2022.

Centola, D., J. Becker, D. Brackbill, and A. Baronchelli. 2018. "Experimental Evidence for Tipping Points in Social Convention." *Science* 360 (6393): 1116–19. https://doi.org/10.1126/science.aas8827.

Cullenward, D., and D. G. Victor. 2020. *Making Climate Policy Work*. John Wiley and Sons.

Dafnomilis, I., H. H. Chen, M. den Elzen, P. Fragkos, U. Chewpreecha, H. van Soest, K. Fragkiadakis, P. Karkatsoulis, L. Paroussos, H.-S. de Boer, V. Daioglou, O. Edelenbosch, B. Kiss-Dobronyi, and D. P. van Vuuren. 2022. "Targeted Green Recovery Measures in a Post-COVID-19 World Enable the Energy Transition." *Frontiers in Climate* 4. https://www.frontiersin.org/articles/10.3389/fclim.2022.840933.

Do, T. N., P. J. Burke, H. N. Nguyen, I. Overland, B. Suryadi, A. Swandaru, and Z. Yurnaidi. 2021. "Vietnam's Solar and Wind Power Success: Policy Implications for the Other ASEAN Countries." *Energy for Sustainable Development* 65: 1–11. https://doi.org/10.1016/j.esd.2021.09.002.

Drews, S., and J. C. J. M. van den Bergh. 2016. "What Explains Public Support for Climate Policies? A Review of Empirical and Experimental Studies." *Climate Policy* 16 (7): 855–76. https://doi.org/10.1080/14693062.2015.1058240.

Eskander, S. M. S. U., and S. Fankhauser. 2020. "Reduction in Greenhouse Gas Emissions from National Climate Legislation." *Nature Climate Change* 10 (8): 750–56. https://doi.org/10.1038/s41558-020-0831-z.

Fay, M., S. Hallegatte, A. Vogt-Schilb, J. Rozenberg, U. Narloch, and T. Kerr. 2015. *Decarbonizing Development: Three Steps to a Zero-Carbon Future*. Washington, DC: World Bank. http://hdl.handle.net/10986/21842.

Fesenfeld, L. P., N. Schmid, R. Finger, A. Mathys, and T. S. Schmidt. 2022. "The Politics of Enabling Tipping Points for Sustainable Development." *One Earth* 5 (10): 1100–08. https://doi.org/10.1016/j.oneear.2022.09.004.

Geels, F. W. 2006. "Major System Change through Stepwise Reconfiguration: A Multi-level Analysis of the Transformation of American Factory Production (1850–1930)." *Technology in Society* 28 (4): 445–76. https://doi.org/10.1016/j.techsoc.2006.09.006.

Geels, F. W. 2012. "A Socio-Technical Analysis of Low-Carbon Transitions: Introducing the Multi-level Perspective into Transport Studies." *Journal of Transport Geography* 24: 471–82. https://doi.org/10.1016/j.jtrangeo.2012.01.021.

Geels, F. W., B. K. Sovacool, T. Schwanen, and S. Sorrell. 2017. "The Socio-Technical Dynamics of Low-Carbon Transitions." *Joule* 1 (3): 463–79. https://doi.org/10.1016/j.joule.2017.09.018.

Hidalgo, C. A., B. Klinger, A. L. Barabási, and R. Hausmann. 2007. "The Product Space Conditions the Development of Nations." *Science* 317 (5837): 482–87. https://doi.org/10.1126/science.1144581.

ICAP (International Carbon Action Partnership). 2023. "China National ETS." Fact sheet, ICAP, https://icapcarbonaction.com/en/ets/china-national-ets.

IEA (International Energy Agency). 2023. *Electricity Market Report 2023*. Paris: IEA. https://www.iea.org/reports/electricity-market-report-2023.

Kelsall, T., N. Schulz, W. D. Ferguson, M. vom Hau, S. Hickey, and B. Levy. 2022. "The Idea of a Political Settlement." In *Political Settlements and Development: Theory, Evidence, Implications*, edited by T. Kelsall, N. Schulz, W. D. Ferguson, M. vom Hau, S. Hickey, and B. Levy. Oxford University Press. https://doi.org/10.1093/oso/9780192848932.003.0002.

Kuss, P., and K. A. Nicholas. 2022. "A Dozen Effective Interventions to Reduce Car Use in European Cities: Lessons Learned from a Meta-Analysis and Transition Management." *Case Studies on Transport Policy* 10 (3): 1494–513. https://doi.org/10.1016/j.cstp.2022.02.001.

Li, L., and A. Taeihagh. 2020. "An In-Depth Analysis of the Evolution of the Policy Mix for the Sustainable Energy Transition in China from 1981 to 2020." *Applied Energy* 263: 114611. https://doi.org/10.1016/j.apenergy.2020.114611.

Lockwood, M. 2015. "The Political Dynamics of Green Transformations: Feedback Effects and Institutional Context." In *The Politics of Green Transformations*, edited by I. Scoones, M. Leach, and P. Newell. Routledge.

Lockwood, M. 2022. "Policy Feedback and Institutional Context in Energy Transitions." *Policy Sciences* 55: 487–507. https://doi.org/10.1007/s11077-022-09467-1.

Mareschi, E., N. Vogel, A. Larsson, S. Perander, and T. Koglin. 2022. "Residents' Acceptance towards Car-Free Street Experiments: Focus on Perceived Quality of Life and Neighborhood Attachment." *Transportation Research Interdisciplinary Perspectives* 14: 100585. https://doi.org/10.1016/j.trip.2022.100585.

Markard, J., A. Rinscheid, and L. Widdel. 2021. "Analyzing Transitions through the Lens of Discourse Networks: Coal Phase-out in Germany." *Environmental Innovation and Societal Transitions* 40: 315–31. https://doi.org/10.1016/j.eist.2021.08.001.

Mealy, P., P. Barbrook-Johnson, M. Ives, S. Srivastav, and C. Hepburn. 2023. "Sensitive Intervention Points: A Strategic Approach to Climate Action." Oxford Working Paper No. 2023-15, Institute for New Economic Thinking, Oxford Martin School, University of Oxford.

Mealy, P., and D. Coyle. 2022. "To Them That Hath: Economic Complexity and Local Industrial Strategy in the UK." *International Tax and Public Finance* 29 (2): 358–77. https://doi.org/10.1007/s10797-021-09667-0.

Mealy, P., M. Ganslmeier, C. Godhino, and S. Hallegatte. Forthcoming. "Climate Policy Feasibility Frontiers: A Tool for Realistic and Strategic Climate Policy Making." Policy Research Working Paper, World Bank, Washington, DC.

Meckling, J., N. Kelsey, E. Biber, and J. Zysman. 2015. "Winning Coalitions for Climate Policy." *Science* 349 (6253): 1170–71. https://doi.org/10.1126/science.aab1336.

Meckling, J., T. Sterner, and G. Wagner. 2017. "Policy Sequencing Toward Decarbonization." *Nature Energy* 2: 918–22. https://doi.org/10.1038/s41560-017-0025-8.

Milkoreit, M. 2022. "Social Tipping Points Everywhere? Patterns and Risks of Overuse." *WIREs Climate Change* 14 (2): e813. https://doi.org/10.1002/wcc.813.

Nacke, L., A. Cherp, and J. Jewell. 2022. "Phases of Fossil Fuel Decline: Diagnostic Framework for Policy Sequencing and Feasible Transition Pathways in Resource Dependent Regions." *Oxford Open Energy* 1: oiac002. https://doi.org/10.1093/ooenergy/oiac002.

Nascimento, L., and N. Höhne. 2023. "Expanding Climate Policy Adoption Improves National Mitigation Efforts." *NPJ Climate Action* 2 (12). https://doi.org/10.1038/s44168-023-00043-8.

Nascimento, L., T. Kuramochi, G. Iacobuta, M. den Elzen, H. Fekete, M. Weishaupt, H. L. van Soest, M. Roelfsema, G. De Vivero-Serrano, S. Lui, F. Hans, M. J. de Villafranca Casas, and N. Höhne. 2021. "Twenty Years of Climate Policy: G20 Coverage and Gaps." *Climate Policy* 22 (2): 158–74. https://www.tandfonline.com/doi/full/10.1080/14693062.2021.1993776.

OECD (Organisation for Economic Co-operation and Development). 2023. *Climate Tipping Points: Insights for Effective Policy Action.* Paris: OECD Publishing. https://www.oecd.org/environment/climate-tipping-points-abc5a69e-en.htm.

Pahle, M., D. Burtraw, C. Flachsland, N. Kelsey, E. Biber, J. Meckling, O. Edenhofer, and J. Zysman. 2018. "Sequencing to Ratchet up Climate Policy Stringency." *Nature Climate Change* 8: 861–67. https://www.nature.com/articles/s41558-018-0287-6

Rentschler, J., and M. Bazilian. 2017. "Reforming Fossil Fuel Subsidies: Drivers, Barriers and the State of Progress." *Climate Policy* 17 (7): 891–914. https://doi.org/10.1080/14693062.2016.1169393.

Rosenbloom, D., J. Meadowcroft, and B. Cashore. 2019. "Stability and Climate Policy? Harnessing Insights on Path Dependence, Policy Feedback and Transition Pathways." *Energy Research and Social Science* 50: 168–78. https://doi.org/10.1016/j.erss.2018.12.009.

Sato, M., R. Rafaty, R. Calel, and M. Grubb, M. 2022. "Allocation, Allocation, Allocation! The Political Economy of the Development of the European Union Emissions Trading System." *WIREs Climate Change* 13 (5): e796. https://doi.org/10.1002/wcc.796.

Sharpe, S., and T. M. Lenton. 2021. "Upward-Scaling Tipping Cascades to Meet Climate Goals: Plausible Grounds for Hope." *Climate Policy* 21 (4): 421–33. https://doi.org/10.1080/14693062.2020.1870097.

Sönke, E., S. M. Constantino, E. U. Weber, C. Efferson, and S. Vogt. 2022. "Group Identities Can Undermine Social Tipping after Intervention." *Nature Human Behaviour* 6 (12): 1669–79. https://doi.org/10.1038/s41562-022-01440-5.

Van Der Ploeg, F., and A. J. Venables. 2022. "Radical Climate Policies." Policy Research Working Paper, World Bank, Washington, DC. https://doi.org/10.1596/1813-9450-10212.

Vogt-Schilb, A., G. Meunier, and S. Hallegatte. 2018. "When Starting with the Most Expensive Option Makes Sense: Optimal Timing, Cost and Sectoral Allocation of Abatement Investment." *Journal of Environmental Economics and Management* 88: 210–33. https://doi.org/10.1016/j.jeem.2017.12.001.

Zaccaria, A., M. Cristelli, A. Tacchella, and L. Pietronero. 2014. "How the Taxonomy of Products Drives the Economic Development of Countries." *PloS One* 9 (12): e113770. https://doi.org/10.1371/journal.pone.0113770.

4 Policy Design

Managing the Distributional Effects of Climate Policies

KEY INSIGHTS

Climate policies have heterogenous distribution implications across income classes, sectors, occupations, and space. The variance in impacts is larger within than across income groups: whether policies are progressive or not, political opposition is more likely to originate from impacts concentrated on sectors or places. Near-poor and lower-middle-class households often experience larger and more visible immediate impacts, making them more likely to oppose policies than poor people, who tend to have limited access to modern energy or transportation.

Compensation to protect poor and vulnerable populations is possible and affordable, but may involve practical challenges related to targeting and delivering mechanisms. Most countries lack the social protection infrastructure and household data to target and deliver compensation.

Some communities face highly concentrated impacts and require specific interventions that go beyond protecting directly affected workers and households. Place-based policies can support the transition of these communities but need to tackle well-identified barriers, combine well-coordinated interventions, and be rigorously evaluated.

Because climate policies can have disproportionate impacts on certain population groups, understanding and addressing these distributional effects is key for navigating political economy challenges—that is, different interests. Ample evidence shows that, in the aggregate, well-designed low-carbon structural change can be a positive change. But sectoral, regional, and household-level outcomes are heterogenous, and the impacts could be damaging for certain population groups unless those groups are actively protected and supported. These distributional impacts shape the interests of population groups and are therefore decisive for support for or opposition to policy reforms.

Community relations also shape the distributional effects of climate policies. As well as analyzing distributional impacts of climate policies for individuals, households, and population groups, it is important to consider the impacts at the community level. Local communities can be empowered to improve the ways they relate to each other, organize themselves, and work together. For example, in Indonesia, local norms and practices, such as *gotong royong*—traditions of collective action, obligations toward others, and mutual assistance—contribute to adaptive capacity (World Bank, forthcoming). A solely household-centric understanding of distributional aspects of climate policies can overlook these community dynamics and codependencies.

Using the example of a tax reform that increases the price of fossil fuels, this chapter highlights that distributional effects are a crucial—albeit not the only—element of the political economy of climate policies. Understanding distributional effects can help countries not only identify and support population groups that are particularly vulnerable to adverse policy impacts but also recognize and manage potential sources of opposition. The chapter highlights the channels through which distributional impacts occur, across income groups, geographical areas, and social vulnerabilities.

Drivers of social exclusion and injustice

Although the focus is often on income groups, occupations, or space, other social factors—such as ethnicity, gender, sex, age, disability, religion, displacement, and sexual orientation—can make groups more vulnerable to the adverse impacts of policy decisions. By evaluating the impacts of policies in their real implementation context, ex post studies offer insights into questions of social justice, intersecting inequalities, and other factors that determine the real impacts of climate policies (box 4.1).

The distinction between *socioeconomic vulnerability* and *social vulnerability*—with the former typically based on income, consumption, or wealth, and the latter on factors that

---

BOX 4.1

Gender and climate change mitigation policies in agriculture

Despite women's important contribution to agriculture, gender inequalities persist and women tend to have less access to resources, including land, inputs, financial services, education, and decent employment opportunities (Erman et al. 2021). Gender roles in agriculture have implications not only for how climate mitigation policies affect social equity outcomes but also for how different groups adjust to these policies and whether they facilitate a just transition for all. If they do not recognize gender dynamics and related transitional challenges, mitigation policies and instruments could exacerbate existing gender inequalities. Technological solutions to mitigate greenhouse gas emissions and reduce hard physical work create opportunities and trade-offs (Beuchelt and Badstue 2013). Long-term gains in reducing emissions, women's drudgery, and unpaid labor do not prevent short-term transitional impacts. Emissions-reducing technologies that lower labor requirements also pose transitional challenges for women workers, who often do the more labor-intensive work in agriculture.

Making mitigation policies gender-responsive poses a challenge because those who can benefit often lack political voice. The lack of power, voice, and recognition is evidenced from the beneficiary level (Larson et al. 2015) to national and international decision-making and governance frameworks (Gautam et al. 2022). A study on gender, power, and decision-making in the Bolivian Amazon (Boyd 2010) reports that, at the beneficiary and community levels, bids to make project design and implementation participatory and inclusive have focused mostly on "practical gender needs" such as health, education, income generation, and food production, rather than on recognizing "strategic gender needs" and women's interests, which can empower them and change their status in society, thus advancing their interests.

(Continued)

BOX 4.1
Gender and climate change mitigation policies in agriculture (continued)

Recognizing women's knowledge and role in sustainable practices and creating meaningful opportunities for decision-making can increase mitigation effectiveness and reduce gender gaps. Many initiatives in Bangladesh, Indonesia, and Nicaragua have empowered and mobilized women in disaster risk management (World Bank 2011). In Bangladesh, engaging women in decision-making and community disaster risk management practices has garnered wider support from both men and women for women's disaster risk management needs, also addressing cultural reasons that prevent women from accessing shelters during disasters (Ikeda 2009). This change has helped reduce the gender gap in disaster-related mortality rates in Bangladesh (World Bank 2011). Women are agents of change who have a strong body of knowledge and expertise through their interactions with the environment. A wide range of literature from Africa (Braun and Duveskog 2011; Friis-Hansen and Duveskog 2012), Asia, and Latin America (Ashby et al. 2000; Humphries et al. 2012) demonstrates that men and women engage with the environment differently and acquire different skills, which can be used in climate mitigation (and adaptation) in agriculture.

Source: Kabir, De Vries Robbe, and Godinho, forthcoming.

make population groups socially vulnerable—is important from a policy perspective. Recent estimates confirm that substantially more people are at risk of being excluded on the basis of identity, circumstances, and socioeconomic considerations than are people living in monetary poverty. For the latest available year (2017), 2.33 billion to 2.43 billion people, or 31.1 percent to 32.4 percent of the global population, are at risk of exclusion (Cuesta López-Noval and Niño-Zarazúa 2022).

Exclusion and other forms of deteriorated social sustainability are multidimensional in nature. A recent analysis of intersecting social vulnerabilities in South Africa estimates that 65 percent of its population is multidimensionally excluded (Ballon and Cuesta, forthcoming). The incidence of multiple exclusions is higher among women (70 percent) than men (58 percent); and, at 74 percent, Black people are more than twice as likely as White people to be excluded (29 percent). Across all ethnic groups, perceptions of unequally administered laws and poor government performance, along with lack of access to public assistance, contribute most to multiple exclusions. Confidence in government institutions, ownership of assets, and quality of housing show the largest gaps in exclusion between Black and White populations.

Distributional effects of climate action

Impacts of climate policies on consumption

Because of different consumption patterns, consumption impacts on households tend to vary across income levels within countries, and the near-poor and lower-middle classes are often particularly vulnerable (Dorband et al. 2019; Köppl and Schratzenstaller 2022; Steckel, Renner, and Missbach 2021). In many lower-income countries, poorer populations have limited access to energy-consuming assets and services—such as cars, air conditioning, or gas for heating and cooking—and are therefore less exposed than richer households to the direct impacts of an increase in fossil fuel prices. Thus, carbon pricing tends to have a progressive effect in countries where poor households have lower-than-average energy expenditure, as is often the case in lower-middle-income countries (figure 4.1). In low-income countries, overall consumption impacts tend to be smaller because households across the whole income spectrum spend relatively less on fossil fuels and energy, but those

FIGURE 4.1. Consumption impacts of a (noncompensated) increase in fuel prices in a subset of countries, by income level

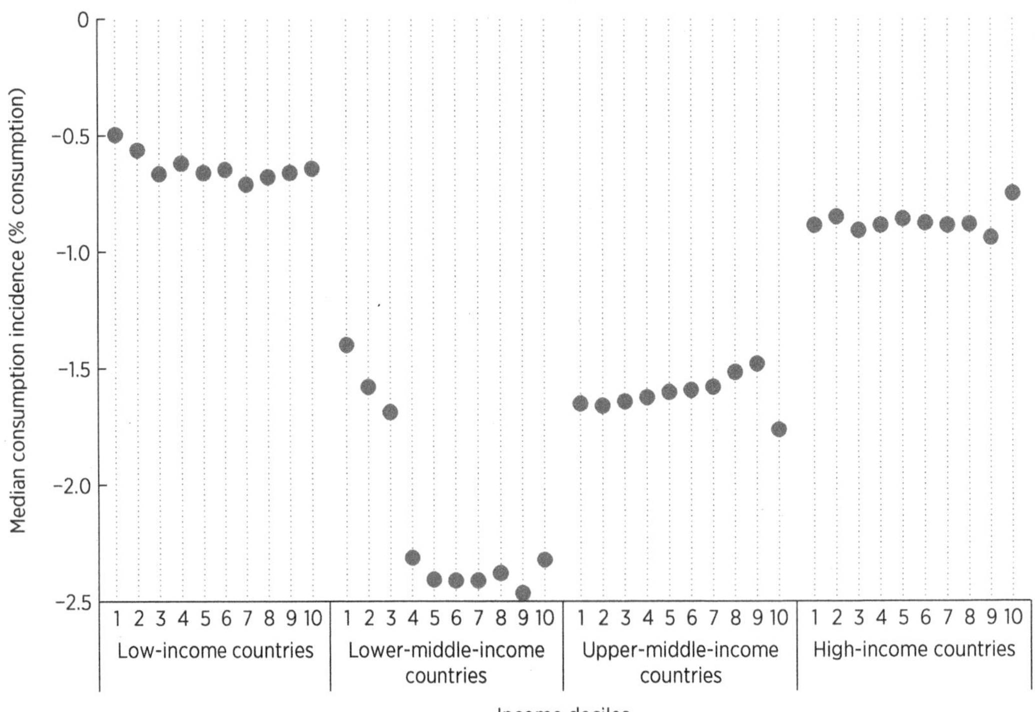

Source: Dorband, forthcoming, using the Climate Policy Assessment Tool developed by the International Monetary Fund and the World Bank to estimate the impact of carbon pricing (https://www.worldbank.org/en/topic/climatechange/brief/climate-policy-assessment-tool).

Note: These illustrative simulations, performed in 74 countries, assume the introduction of a tax of US$60 per ton of carbon dioxide and the removal of energy and fossil fuel subsidies, with no recycling of the revenues or savings. These assumptions are meant not to be realistic policy packages but to illustrate the vulnerability of households to changes in fuel prices. The figure shows median effects for consumption deciles and for country groups. It does not show the large heterogeneity of consumption effects across countries as well as within countries or consumption deciles. Because impacts are heterogenous and skewed, they are larger if averages are used instead of medians, but the distribution does not change substantially.

impacts tend to be neutral or slightly progressive. Among upper-middle-income countries, the evidence is more mixed and varies with levels of access to public transportation and electricity and other lower-carbon alternatives. In high-income countries, where most households have access to energy-consuming assets, lower-income households are slightly more vulnerable to energy price increases because they spend a larger portion of their expenditure on fuels. Irrespective of the distribution across income groups, a 1 percent decrease in consumption is expected to have larger adverse effects on the welfare of relatively poorer individuals. This creates a specific vulnerability for the near-poor and lower-middle classes, who have access and enough resources to consume fossil fuels but are vulnerable to small changes in price and reductions in their purchasing power.

Poor people tend to spend a large fraction of their income on food and can therefore be heavily affected if climate policies translate into higher food prices. Agricultural products and their supply chains are major greenhouse gas emitters, through agriculture's use of fertilizers, transportation, and fuels, and through deforestation (WRI 2019). Although carbon pricing systems rarely cover land-based emissions, climate policies that affect agriculture and food systems could have large impacts on food prices.

For example, in Bolivia, 70 percent of the impact of carbon pricing on the consumption of the bottom quintile would come from food prices (Vogt-Schilb et al. 2019). And that estimate assumes that farmers can pass the increased cost of inputs to final food prices. If, instead, farmers must reduce their use of agricultural inputs, agricultural yields could be reduced and the resulting increase in food prices could hurt poor people even more. Carbon pricing with a narrower base—covering only energy-related emissions, as is most common—has much smaller impacts on food prices in most countries, preventing such a regressive impact (Dorband et al. 2019).

Available estimates on distributional outcomes across low-, middle-, and high-income economies are inconclusive, and depend on the methods used and policy assumptions made (Ohlendorf et al. 2021). Some studies find progressive impacts of carbon pricing in China (Brenner, Riddle, and Boyce 2007), India (Datta 2010), Indonesia (Steckel et al. 2021; Yusuf 2008), Mexico (Renner 2018), Nigeria (Dorband et al. 2022), Pakistan (Shah and Larsen 1992), South Africa (van Heerden et al. 2005), Thailand (Saelim 2019), and Viet Nam (Nurdianto and Resosudarmo 2016; Steckel et al. 2021). Others find neutral or regressive distributions in Bolivia and Ethiopia (Steckel, Renner, and Missbach 2021); Brazil (da Silva Freitas et al. 2016); China (World Bank Group 2022b); India (Steckel et al. 2021); Indonesia, Malaysia, and the Philippines (Nurdianto and Resosudarmo 2016); and South Africa (Devarajan et al. 2011; Steckel, Renner, and Missbach 2021). In higher-income countries, regressive consumption impacts are more likely (Beck et al. 2015; Dissou and Siddiqui 2014; Feng et al. 2010; Goulder et al. 2019; Grainger and Kolstad 2010; Kerkhof et al. 2008; Kerkhof, Nonhebel, and Moll 2009; Wier et al. 2005). More recent research across European countries, however, finds neutral or progressive consumption effects in most countries (Andersson and Atkinson 2020; Feindt et al. 2021).

Consumption incidence studies by income group do not capture all distributional issues—and related political economy risks—because differences *within* income groups are larger than variations *across* income groups (Dorband et al. 2022; Douenne 2020; Feindt et al. 2021; Missbach, Steckel, and Vogt-Schilb 2022; Steckel, Renner, and Missbach 2021). Consumption patterns vary more with socioeconomic characteristics that are unrelated to income, such as access to clean energy types or transportation modes (Javaid, Creutzig, and Bamberg 2020; Malakar, Greig, and van de Fliert 2018; Muller and Yan 2018). Differences also occur between rural and urban settings (Dorband et al. 2022; Douenne 2020; Feindt et al. 2021). For example, estimates suggest that, in both low- and high-income countries, low-income rural households pay a greater budget share for carbon pricing than do their urban peers. Often with limited access to public transportation and electricity, especially in low-income countries, these households tend to spend a large share of their budget and time on acquiring cooking and transportation fuels, including for electricity generators. In many countries, particularly in Sub-Saharan Africa, households spend more on generator fuel than on grid electricity, and generators produce more electricity annually than the national grid (IFC 2019). Because transportation fuels are also often subsidized, a pricing reform that includes the removal of these subsidies may particularly affect low-income rural households.

In low- and middle-income countries, higher fuel prices may not immediately affect poor people; without complementary action and access to affordable electricity, however, such policies could slow progress toward universal access to modern energy and clean cooking. Consumption incidence studies that take a static view—that is, looking at today's consumption only—may fail to identify such long-term risks. Thus, it is important that assessments of climate policies and their distributional impacts be carried out in a dynamic fashion, particularly in rapidly growing countries. For example, an increase in fossil fuel prices may not directly affect households cooking with biomass, but the change in price may delay the ability of these households to shift to modern cooking techniques if they do not have access to affordable electricity and electric cookstoves (Greve and Lay 2023). In 2020, approximately 2.4 billion people around the world cooked with traditional polluting fuels and technologies, contributing to air pollution and premature deaths (World Bank 2023a). Switching to cleaner cooking fuels and modern techniques is a necessary step to improve people's health and well-being. But the switch might lead to an important expenditure factor if fuel prices increase, especially in areas with little access to electricity and where biomass is available free of charge. Carbon pricing might therefore reduce the uptake of modern cooking fuels among poor households that cannot afford the transition (Rao 2015). Cameron et al. (2016) provide evidence that climate change mitigation policies in South Asia could increase fuel costs by 38 percent in 2030 relative to a baseline scenario, and risk keeping 21 percent more people on traditional stoves.

Investment, tax reforms, and cash transfers can protect people against the direct impacts on consumption

Climate policies coupled with infrastructure investments can be strongly progressive, but complementary policies—such as immediate monetary transfers—are often necessary in the short term. Not only is access to electricity crucial for improving well-being and productivity, but it can also help insulate households from fossil fuel price shocks, provide cleaner and safer energy alternatives, and support poverty reduction (see, for example, Fagbemi, Osinubi, and Adeosun 2022). Because lack of access to basic public services is concentrated among low-income individuals and rural communities (Dorband et al. 2022), investing carbon price revenues in improving energy access would particularly benefit poor and rural groups. But infrastructure provision and its associated welfare benefits take time to materialize and cannot compensate for the immediate consumption shock. Thus, although cash transfers cannot replace more structural reforms toward achieving the Sustainable Development Goals, immediate redistribution and compensation will often be needed to alleviate the initial consumption effect and manage public resistance (Boyce 2018; Goulder, Hafstead, and Williams 2016; Hassan and Prichard 2016; Klenert et al. 2018; Metcalf 2008).

Recycling just a fraction of carbon pricing revenues or repurposing subsidies through direct transfer can make reforms pro-poor. Fossil fuel subsidies are often implemented to support poor people's energy access, and carbon pricing is often contested for its impact on energy access. However, because rich households consume the bulk of fuels in absolute terms, energy or fossil fuel subsidy schemes represent an extremely inefficient tool for pro-poor support: most of the resources flow toward higher-income people. These schemes also disincentivize energy conservation efforts. Social safety nets, such as targeted or untargeted cash transfers, by contrast, are more effective and less costly (Banerjee, Niehaus, and Suri 2019; Coady and Le 2020; Hanna and Olken 2018). When countries reduce fossil fuel subsidies or implement carbon pricing, just a fraction of the resources they mobilize is enough to fully compensate the bottom shares of the income distribution. In Ecuador, for example, Schaffitzel et al. (2020) find that removing all energy subsidies and repurposing a share of this revenue to increase the cash transfer program, Bono de Desarrollo Humano, would increase the poorest quintile's real income by 10 percent and leave more than US$1.3 billion for the public budget. For a sample of

Latin American countries, Feng et al. (2018) estimate that about 20 percent of savings from subsidy reforms would fully alleviate the consumption impact on the bottom 40 percent of the income distribution.

Compensating people is hard: it requires appropriate systems and delivery mechanisms, including broad, strong, and flexible social protection systems. The large heterogeneity in impacts makes it difficult to target transfers to support the most affected and vulnerable households. Countries already equipped with high-coverage social protection systems can use these to help households manage price shocks in ways that are both better targeted and more efficient than subsidies. But even the best social protection systems have imperfect coverage and targeting. A recent Latin American study explores the gaps and overlaps between the 20 percent of households most affected by a carbon price, the poorest 20 percent of households, and households that are not covered by social transfers (Missbach, Steckel, and Vogt-Schilb 2022). Highlighting the challenge governments face when compensating households, that study finds that, first, the impacts of carbon pricing are more heterogenous within than across income classes and, second, delivery mechanisms are not always available. For example, the authors find that, in Chile, Colombia, Ecuador, and Uruguay, social transfers do not cover 3–4 percent of households among the poorest 20 percent and most affected by carbon pricing. Some countries, including Indonesia and the Islamic Republic of Iran, have used fossil fuel subsidy reform to finance the creation of new systems to compensate households more efficiently than existing tools (Damania et al. 2023).

As discussed in chapter 5, even redistributive, low-carbon, and fiscally responsible policies to remove fuel subsidies can lead to social unrest, so ensuring participative decision-making and developing a good communication strategy to accompany climate policies is vital. When the price of liquefied petroleum gas—used mostly for vehicles—doubled overnight in Kazakhstan in 2022 after the government lifted price caps, protesters took to the streets and the ensuing turmoil led to more than 200 deaths and the resignation of the government (Horowitz 2022).

Sectoral effects on labor and skills

Ample evidence shows that well-designed climate policies can be net job creators (Dussaux 2020; Godinho 2022; Markandya et al. 2016; Metcalf and Stock 2020; World Bank Group 2022a). Transitioning toward renewable energy sources and adopting sustainable land use practices can generate new job opportunities in emerging industries. Many studies find that indirect and induced jobs are a major driver of these net-positive outcomes. For example, employment opportunities related to investments in infrastructure, induced by projects in other sectors, can be larger than direct job creation on project sites (Edwards, Sutton-Grier, and Coyle 2013; Zhang and Zhang 2020).

Positive employment outcomes are not automatic; rather, they depend on the design of climate and other policies. Job creation is linked to the transition toward greener technologies and practices—for example, when climate-smart agriculture practices or renewable energy solutions are more job-intensive than existing patterns. But the reality of these new jobs will depend on the investment climate, the availability of infrastructure and a trained labor force, and appropriate tax policies and trade regulations. To become part of a green global value chain, and benefit from the job creation it can bring, countries must be able to import key upstream components at an acceptable cost, which ill-designed tariffs or trade regulations may prevent them from doing.

To understand the effect of climate and development policies on labor and skills, an analysis of climate policy packages that promote sustainable development was conducted using a global demand-led economic model based on multiregional input-output data (Dorband, forthcoming). The policy scenario is a US$60 domestic carbon tax accompanied by a complete phase-out of fossil fuel subsidies, with generated revenues split as follows: 40 percent on income support (income tax cuts or social safety nets); 40 percent on government spending related to the Sustainable Development Goals (education, health, staple crop production, and public transportation); and, where applicable, 20 percent on renewable wind and solar energy subsidies. The analysis included the 121 countries in the Global Trade Analysis Project (Aguiar et al. 2019).

The analysis finds that, despite mostly positive net effects, the climate policy packages can result in sizable sectoral job reallocations and policy-induced structural changes, particularly in more carbon-intensive economies. Figure 4.2 shows the effects of these packages on job losses (y-axis) and gains (x-axis). Countries above the diagonal have net gains in jobs, and those below experience net losses. Even countries with net gains can experience job losses as large as 1 percent or 2 percent of total employment, whereas gains amount to between 3 percent and 4 percent. Although carbon-intensive economies may not experience larger net effects, they do undergo larger structural changes, with greater reshuffling of jobs.

Social protection and active labor policies can reduce and help manage concentrated sector impacts

Countries may need to provide additional social and locational support to facilitate the labor market and skills transition and to reduce potential social and economic frictional costs. Reallocating workers to other sectors will have distributional and equity effects across income groups, skill levels, and occupations, as well as between provinces and countries (Azevedo, Wolff, and Yamazaki 2018; Hille and Möbius 2019; Marin and Vona 2019; Yamazaki 2017). The nature and quality of compensation, stability, protection, and occupational safety offered may also vary between lost and new jobs (Botta 2019). As such, governments may need to put targeted social, labor market, or locational support measures in place to facilitate job transitions, ease skill mismatches and frictional costs, and increase public acceptability (Saussay et al. 2022; Vona 2019).

Barriers to labor market mobility can significantly increase unemployment outcomes associated with low-carbon structural change. For example, in the United States, Castellanos and Heutel (2019) show that the modeled unemployment impacts of a carbon tax were 24 percent higher when the labor market was assumed to be perfectly immobile—that is, workers could not easily change jobs—as opposed to being perfectly mobile—that is, when workers could frictionlessly transition into any job, even in a different location. Similarly, in Brazil, Berryman et al. (forthcoming) show that accounting for empirically derived patterns of job switching across occupations and geographies significantly increases unemployment outcomes associated with modeled scenarios intended to boost productivity and reduce deforestation. They also find that, in these scenarios, workers in the bottom income deciles, particularly those in the agriculture sector, are likely to see the most adverse unemployment impacts.

Countries need to make sure they target policies to reduce adverse employment outcomes and increase labor market flexibility. For example, coal miners have skills that

FIGURE 4.2. Effects of climate policy packages on job losses and gains in 121 countries, by carbon intensity of countries' economies

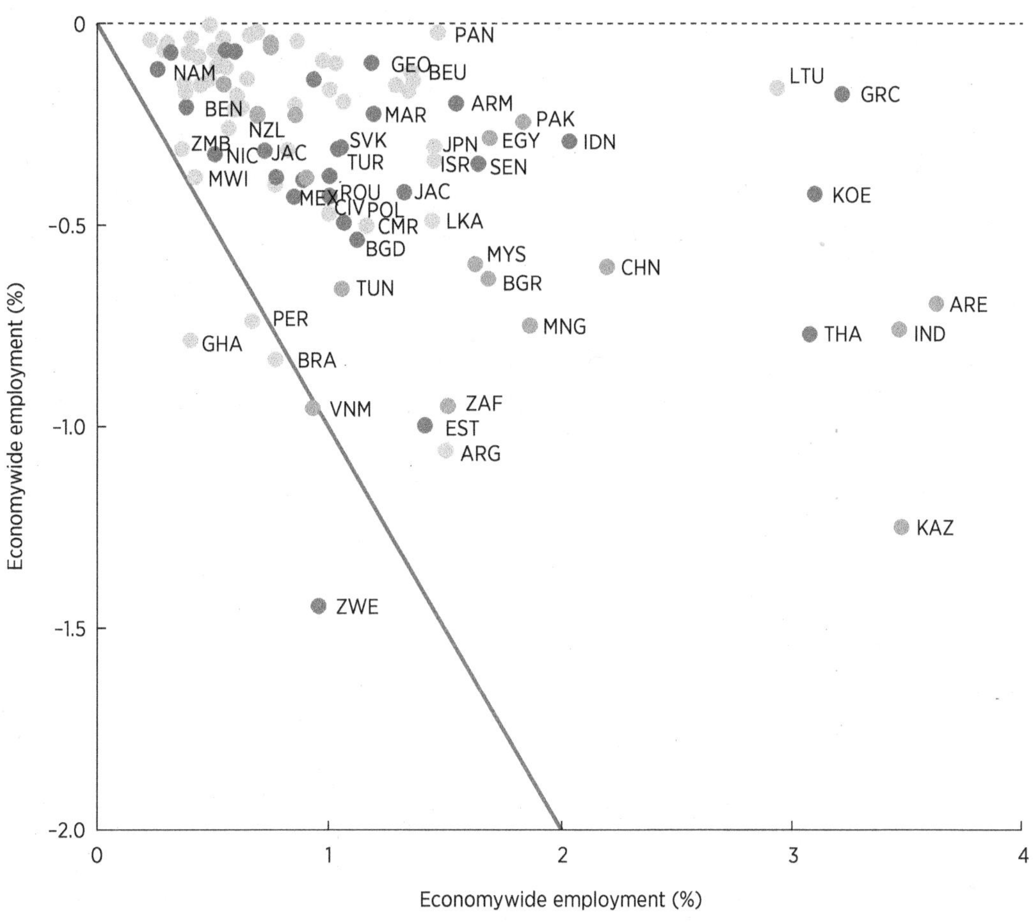

Source: Dorband, forthcoming, using the MINDSET model.
Note: Net labor demand increases in countries to the right of the green line. Light blue dots show countries with high carbon intensity, gray dots show countries with intermediate carbon intensity, and yellow dots show countries with low carbon intensity.

could be adapted to mining green minerals, such as graphite, but would have to relocate because these mines are in different regions. Similarly, petroleum engineers have skills that could allow them to transition into several alternative occupations, but many of these occupations do not pay as well. For active labor market policies to effectively help people transition, governments must therefore ensure they target the policies at the constraints workers face. Comprehensive adaptive support needs to take a broad approach, using initiatives such as counseling and other social services for workers and their families and supporting their reentry into jobs (box 4.2).

Empirical studies on the employment impacts of climate policies highlight how political economy factors also shape employment outcomes (Godinho 2022). Model-based assessments of the impacts of energy and sustainability transitions may not capture the way in which political economy factors relating to institutional structures, power asymmetries, and other contextualities can shape policy outcomes (Somanathan et al. 2014). Empirical studies offer an important complement by reflecting sociopolitical, labor

BOX 4.2

Achieving a just transition in agriculture

Given the large population shares employed in the sector, especially in South Asia and Sub-Saharan Africa, understanding the nature of agricultural employment, and how and what workers will transition into when mitigation policies are implemented, is imperative. Case studies from Brazil and Mali show that skill and spatial mismatches will likely make it difficult for workers to transition into jobs outside agriculture. Despite the possibility of transitions into nonagricultural occupations, historically these transitions are rare and require a high degree of retraining, skills upgrading, and, in the long term, economic diversification. The seasonal, and often transient, nature of the agricultural workforce further complicates achieving a just transition in agriculture. Agricultural mitigation policies affect employment and wages through total factor productivity and value-added inputs (Dorin, Hourcade, and Benoit-Cattin 2013). If policies reduce productivity, agriculture will require more labor and land, slowing transformation; however, boosting productivity—for example, through climate-smart practices—can lead to long-term labor release (Fuglie et al. 2020; Gautam et al. 2022). As labor market challenges hinder a quick transition, policies that boost productivity risk creating unemployment for workers who are already highly vulnerable.

A just transition does not mean transitioning out of agriculture. Depending on the structure and development of the agriculture sector, a country's green transition in agriculture may focus on

- Helping (primary) agricultural workers adapt and become more resilient to the effects of climate change by adopting and building new skills for climate-smart practices;
- Exploring avenues to help workers transition out of primary agriculture production to activities within the agriculture value chain;
- Structurally transforming the country's economy by building know-how and diversifying into new products to create opportunities outside of agriculture;
- Raising agricultural productivity by, for example, investing in research and dissemination (Fuglie et al. 2020; Fuglie et al. 2020, 2022);
- Repurposing agricultural subsidies, for example, toward safety nets and to farmers' income (Gautam et al. 2022); and
- Helping workers upgrade their skills for productivity-enhancing tasks and transition into other jobs (Townsend et al. 2017).

Justice concerns should be at the center of any sustainability efforts, not only for farmers as primary producers and land managers but also for farm and food chain workers, consumers, rural communities, and other marginalized groups (Baldock and Buckwell 2022). Ensuring inclusivity and distributional equity, and recognizing the voice, values, and rights of those who work in and consume agriculture, is key to achieving environmental and food security goals while ensuring climate justice for those whose livelihoods depend on agriculture and those who consume its products.

market, and other factors that determine realized outcomes—specifically, who wins or loses. For example, several studies on renewable energy policies highlight that, because of immigration or offshoring, job creation does not always benefit local populations (del Río and Burguillo 2009; Jumani et al. 2017; Obour et al. 2016). And, even with job creation, workers may not feel like winners when they have low wages, safety, or security (Cormack and Kurewa 2018; Huesca-Pérez, Sheinbaum-Pardo, and Köppel 2016; Terrapon-Pfaff et al. 2019), or when employment outcomes are short-lived or decrease after the initial construction period (Cai et al. 2016; Leistritz and Coon 2009; Ortega et al. 2015). The political economy can also affect who benefits, with studies from renewable energy projects in India and Kenya revealing that political affiliation or ethnicity may influence who gets jobs (Cormack and Kurewa 2018; Lakhanpal 2019). To correct such unfair distributional impacts and support a just transition, countries may therefore need to take additional policy approaches, based on engagement with affected communities and workers.

Although governments can use cash transfers to compensate for monetary losses—for example, losses due to changes in relative prices—it is harder to compensate for the loss of employment and livelihoods, and even more so for losing culture or a sense of place. Thus, viewing job losses solely from an income perspective misses some key aspects, such as the sense of community derived from work, family structures based on the division of household and external tasks, or the identity given by an aspect of labor. For example, US estimates of the median subsidy required for a person to be indifferent about moving to a similar location exceed 100 percent of annual income, increasing by 43 percent if the person has family living in the original location (Bartik 2020). These high numbers explain the relatively limited migration across regions and the persistence of local underemployment hot spots, even in countries with a low level of aggregate unemployment.

Green industrial policies can build political support and reduce the cost of green technology

As well as fostering winners, green industrial policy can help reduce impacts for potential policy losers and smooth the transition. Rozenberg, Vogt-Schilb, and Hallegatte (2020) demonstrate how enacting regulations that apply only to new capital, as fuel economy standards or "feebate" programs do in the automobile sector, can favor a transition toward a greener economy without negatively affecting those who depend on existing polluting capital. Avoiding impacts on the value or price of existing assets certainly helps overcome political economy barriers, because it prevents impacts that are often concentrated and considered unfair. But acting only on new investments without changing the incentives regarding the use of assets makes the transition slower, creating a trade-off between its pace and political acceptability.

Green industrial policies can support the effectiveness of other policies and reduce their distributional impacts but are not necessarily a substitute for other approaches. Hallegatte, Fay, and Vogt-Schilb (2013) developed a simple matrix to explore the role of, and relationship between, green industrial and pricing policies in the climate policy toolkit (figure 4.3). The matrix is based on two interdependent factors: *price effectiveness* (the extent to which pricing instruments can trigger the needed structural changes) and *price adequacy* (the extent to which the political economy makes it possible to adjust price levels to the level needed to change behaviors). Both depend on elasticity of demand and influence which policies will be feasible and efficient.

- If demand for a good or service is very responsive to even small changes in price—such as dispatch decisions in the power system—effective prices can be quite low, which tends to be socially and politically acceptable (quadrant 1 in figure 4.3). In this case, price-based mechanisms, or regulations, can be efficiently implemented.
- When changes in price have little effect on demand, as is the case with household decisions regarding transportation or decisions to influence long-term research and development for steel production, the price needed to change behaviors and technologies would need to be very high. That high price would in turn likely trigger pushback from affected groups. In that case, it makes sense to start by implementing policies that increase price elasticity by creating substitution options, such as green industrial

FIGURE 4.3. **A matrix to determine when and how to deploy green industrial policies**

		Effectiveness of price instruments (market failures)	
		High	**Low**
Adequacy of prices (political economy and government failures)	**High**	(1) Pricing solutions, with appropriate complementary policies to manage distributional effects	(2) Temporary (sunrise) industrial policy to help develop new sectors and technologies
	Low	(3) Policies to smooth transition and reduce competitiveness concerns, such as support for sunset industries and social or trade policies	(4) Sunrise green industrial policies to help develop green sectors and technologies, alongside efforts to improve adequacy of pricing, such as support for sunset industries and social or trade policies

Source: Adapted from Hallegatte, Fay, and Vogt-Schilb 2013.
Note: Sunrise policies support and accelerate sectors or technologies that are expected to grow in productivity and competitiveness over time and benefit from climate and industrial policies; *sunset policies* smooth the downscaling of declining sectors to minimize transition and social costs.

policies (quadrants 2 and 4). Green industrial policies increase price elasticity by, for example, expanding the availability of substitutes through support for green innovation or industries that could scale green technologies, such as electric vehicle battery manufacturing. Thus, when the government subsequently implements pricing policies and increases the price of fossil fuel cars or coal power, alternatives, such as electric vehicles and renewable energy, are readily available.

- Even when prices are efficient, increasing them might be impossible for political reasons. This may be the case, for example, in places where large fossil fuel subsidies have led to very low energy efficiency and where any change in energy could have large social and economic impacts. In those cases (quadrant 3 or 4), governments can use green industrial policies to reduce the economy's vulnerability to higher prices and transform the political economy. And, when the main political economy obstacle is the political economy of concentrated impacts, governments can use green industrial policies to support sunset industries to facilitate their downscaling or adjustment.

To reap the full benefits of green industrial policies, countries need to carefully manage some political economy risks, including corruption, policy capture, and distributional conflicts. Corruption can increase costs and reduce trust, eroding support for more ambitious climate policy and potential economic gains. It can also lead to policy lobbying or capture, working against adaptive reforms that can improve emissions and economic outcomes, or to a lock-in to certain technologies at the expense of cheaper or more efficient ones (box 4.3). For example, studies from the European Union show that renewable energy and gas lobby groups have recently formed coalitions that accelerated the transition away from coal but could ultimately work against full power sector decarbonization (Lindberg and Kammermann 2021).

The urban dimensions of climate policy impacts

Urban transportation costs constitute a key factor in determining people's access to jobs and public services, including health care and education, as well as people's location and transportation choices. Policies that affect urban accessibility can have impacts on

BOX 4.3

Green industrial policies: How to minimize the risk of capture

It is difficult to anticipate the potential of new technology or a country's latent comparative advantage, and being wrong can have a large cost. As mentioned in Rodrik (2014), industrial policy aims to discover and develop the appropriate new technologies and products and cannot be expected to succeed in all cases. Therefore, a real potential exists for costly failure and waste of scarce public resources, with a real risk that the public will share the cost of failures and small groups capture any benefits.

Green industrial policies face significant risks of capture and rent-seeking behaviors. For this reason, Johnson, Altenburg, and Schmitz (2014) and Pegels (2014) frame the debate on green industrial policy in terms of managing the rent (risk-adjusted above-average profits) created by industrial policy to incentivize investment in green sectors and technologies. The aim of a green industrial policy is to create the appropriate level of rent from green investment to facilitate the green transition.

Rent-seeking behavior is likely to influence policies, even in countries with high institutional capacity and appropriate checks and balances (Anthoff and Hahn 2010; Helm 2010). Neven and Röller (2000) identify sharply partisan political systems, weak governments, and an absence of transparency as factors that increase the likelihood of such problems. Rent capture remains possible, even in the most efficient, balanced, and transparent economy, because industrial lobbies are powerful actors in any economy. When devising and implementing green industrial policy, adhering to three key design principles can help reduce the risk of capture (Alternburg and Assmann 2017; Rodrik 2014):

1. *Embeddedness.* Policy makers should work closely with the private sector to understand how key industries function and how specific bottlenecks hamper growth. Because these factors are highly context specific and can evolve over time, industrial policy should be seen as a joint explorative process whereby public and private entities constantly adapt and collaborate for industrial development.
2. *Discipline.* A close relationship between the government and private sector can pose greater risk of collusion and political capture. Governments should therefore have clear objectives with measurable indicators, routinely monitor firm and program performance, and have the autonomy to change or withdraw incentive packages without being swayed by lobbyists' pressures. As highlighted by Juhász, Lane, and Rodrik (2023), the success of industrial policy is often less about the government's ability to "pick winners" and more about its ability to "let losers go." Clear separation of policy roles, competitive service provisioning, and transparent guidelines can help deter undue influence and maintain integrity.
3. *Accountability.* Ensuring that policy makers are held responsible for industrial policies is crucial. Implementing strict reporting requirements, disclosure obligations, and democratic oversight by central auditing authorities, political parties, courts, and the media can help foster transparency and credibility.

household well-being, because urban transportation services are determinants of labor market outcomes, urban area productivity, and locational options available to households. Significant empirical evidence demonstrates that increased accessibility leads to better individual labor market outcomes, such as reduced unemployment, better-paid jobs, or more formal and more permanent employment (Aslund, Osth, and Zenou 2010; Franklin 2018; Jin and Paulsen 2017). Conversely, climate policies can affect employment accessibility, labor market outcomes, and overall urban productivity. In particular, increased transportation costs could reduce the welfare of the poorest by constraining them to live in locations where housing costs are high or by locking them out of the urban labor market.

In low- and middle-income countries, fuel price increases appear to cause limited accessibility reductions for lower-income households, generally because many of them cannot afford motorized transportation in the first place. A background study for this book investigates the impact of fuel price increases on accessing employment by public transportation in Kinshasa, Democratic Republic of Congo, and Rio de Janeiro, Brazil; it finds that, if households cannot afford public transportation services before the fare change, an increase in fuel prices will seem not to affect them (Nell et al. 2023).

In both cities, a doubling of fuel prices affects the accessibility of middle- and high-income groups the most. Map 4.1 shows losses in absolute accessibility (share of jobs no longer accessible) in Kinshasa and Rio de Janeiro, spatially and by income decile in a scenario with a 100 percent increase in fuel prices and a 75-minute maximum travel time. Overall, the map suggests that fuel price increases do not have regressive impacts in Kinshasa, although many severely affected outliers appear in deciles 1 and 2. Individuals already priced out of using transit would also need to spend an even higher share of their budget to afford public transportation services under the carbon pricing policy, limiting their participation in the urban economy and reducing their labor market prospects (Franklin 2018). This effect is similar to the mechanisms described earlier in this chapter whereby increases in fossil fuel prices can cause households to revert to cooking with biomass, exacerbating the impacts of indoor air pollution on health. Combined with the lack of voice and influence of the poorest households, this higher vulnerability of near-poor households explains why protecting the poorest alone has failed to ensure the wider acceptability of climate or energy policies.

Electrifying public transportation and having compact urban areas can cushion against a loss of accessibility from fuel price increases. In Rio de Janeiro, such increases had relatively little effect on communities near the rail and metro systems because they can rely on decarbonized transportation systems to reach jobs. In Kinshasa, the city's compactness means that accessibility remains high, because distances between jobs and residents are short and require few motorized transportation legs and expenses. This finding is a testimony to the power of compact dense urban areas in connecting workers with job opportunities and highlights the power of land use policies and transportation decarbonization in cushioning the social impacts of fuel pricing and climate policies (Gusdorf and Hallegatte 2007).

Another study for this book estimates the impacts of a 20 percent fuel price increase on households' economic welfare across four income groups in Cape Town, South Africa, capturing the interplay between transportation and housing costs. It accounts for households' dynamic adaptation strategies, highlighting the impact of fuel price increases on inequality. Liotta et al. (2022) find evident spatial inequalities, even within income groups, immediately after introduction of the policy, before households can implement any adaptation strategies (map 4.2). For a given income class, the workers living far from employment centers are more affected by the fuel price increase than those living close to the centers, because of distance and modes of transportation chosen. In the short term, the fuel tax can affect two households in the same income class in largely different ways. Considering aggregated statistics per income class only would hide these effects, showing the importance of a spatial analysis to understand distributional analysis and anticipate political opposition.

Although households in every income group have options to mitigate the adverse impact of fuel price increases on their well-being, the poorest households face the greatest welfare losses. The change in transportation costs, triggered by fuel price increases, will affect the housing and land markets as the attractiveness of locations change. In turn, households can, if their financial position allows them, choose to switch transportation mode when an alternative is available, change employment or housing location, and

MAP 4.1. Losses in accessibility of jobs using public transportation in Kinshasa and Rio de Janeiro, 100 percent fuel price increase scenario with a 75-minute maximum travel time, by area and income decile

a. Kinshasa, Democratic Republic of Congo

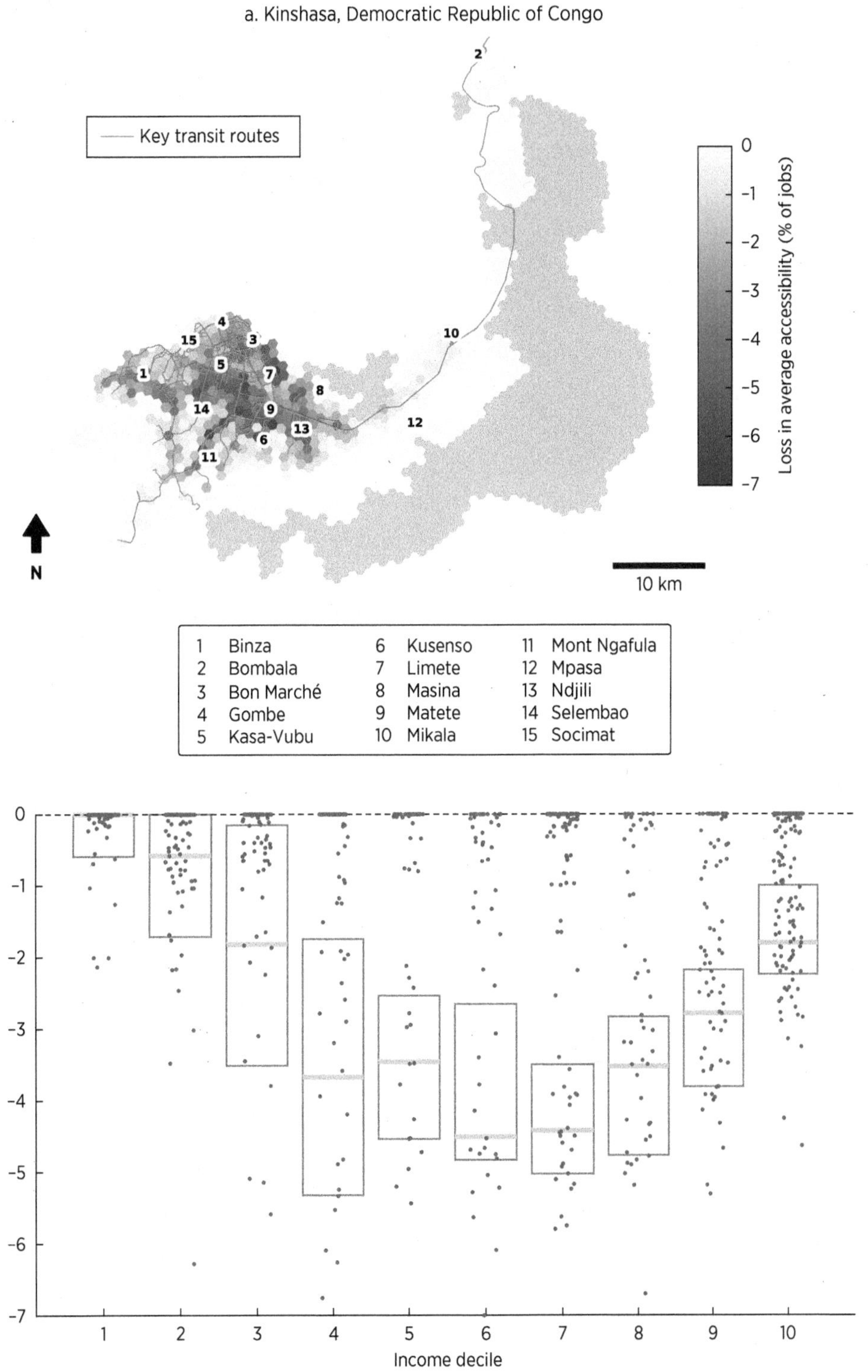

(Continued)

MAP 4.1. Losses in accessibility of jobs using public transportation in Kinshasa and Rio de Janeiro, 100 percent fuel price increase scenario with a 75-minute maximum travel time, by area and income decile (continued)

b. Rio de Janeiro, Brazil

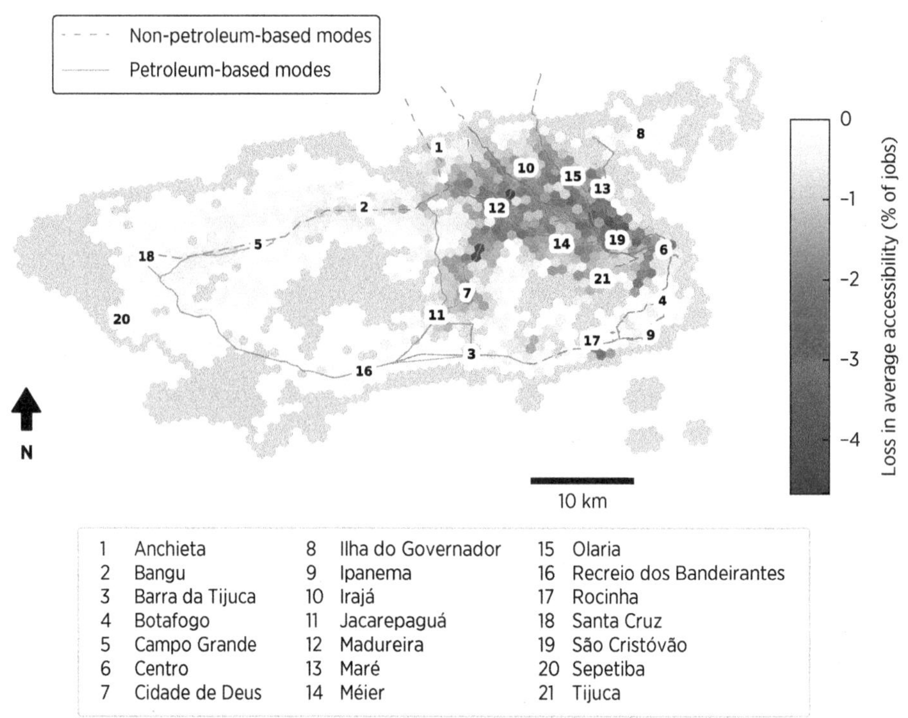

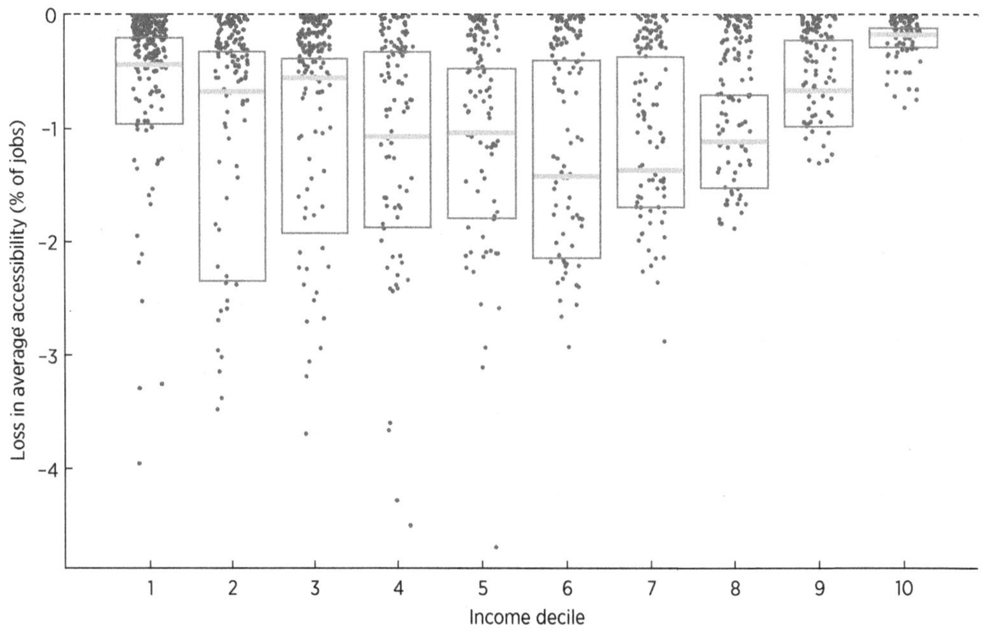

Source: Nell et al. 2023.

MAP 4.2. Direct impact of the fuel tax on incomes net of transportation costs (inclusive of the cost of time) under a scenario that taxes all polluting modes, Cape Town, South Africa

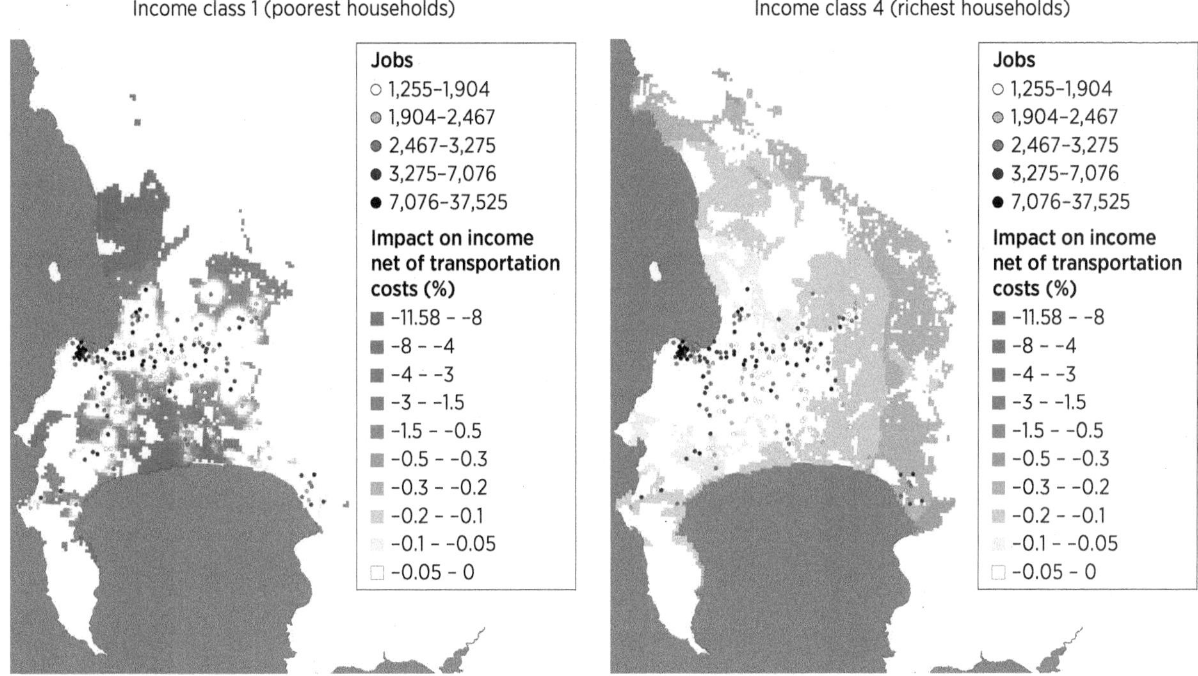

Source: Liotta et al. 2022.

decide to live in smaller or larger dwellings. To some extent, these adaptation strategies and housing market adjustments can limit the impacts of the fuel price increase on households' welfare. The richest households (income groups 3 and 4) experience the least impact, whereas the poorest households suffer the largest welfare losses (income groups 1 and 2).

Urban and transportation policies can mitigate the spatial impacts of climate policies

Governments can mitigate the negative impacts of fuel price increases by improving land and housing policies and regulations and through urban planning. Land and housing markets play a key role in limiting the poorest households' ability to adapt to the transportation cost increases. In Cape Town, subsidized housing and informal dwellings are the only options available for the poorest households, but these options are geographically constrained. Subsidized housing units provided free by local authorities to poor households have positive welfare impacts by freeing up beneficiaries' budget for other priorities; however, their location far from employment centers on average imposes high commuting costs and increases the vulnerability of the poor to transportation cost shocks. Similarly, informal settlements can be erected only on vacant and publicly owned land, generally in the city's outskirts. These geographical constraints leave the poorest households with fewer options for mitigating the fuel price increase—for example, they cannot relocate closer to employment centers. Therefore, reducing housing market rigidities can give them greater flexibility when adapting to changes in the transit system.

Providing access to collective urban transportation can trigger climate benefits; and, because they give rise to fewer concerns about distributional impacts than carbon pricing or fuel taxation, such policies can be more popular than taxation schemes (Carattini, Carvalho, and Fankhauser 2018). Two main types of public transportation policy exist: investing in public transportation infrastructure and operations, and subsidizing public transportation.

- Investments in public transportation infrastructure carry potential large welfare gains, because they allow commuters to save time and money, reduce congestion (with benefits beyond public transportation users), and improve air quality. But such place-based policies can also lead to gentrification, which increases housing rents and displaces poor-income households, giving rise to inequality concerns when the displaced end up worse off or unable to capture the benefits through land or housing prices. Poor renters are particularly affected: they can no longer afford to live in the gentrified area, whereas homeowners can choose to sell, thus benefiting from their property price appreciation.
- Public transportation subsidies can have strong positive welfare impacts but are less efficient and more regressive when they do not target the poor. However, a subsidy scheme that reduces fares by a fixed rate would most benefit those who live far from their jobs. Evidence from the United States shows that public transportation subsidies benefit those living in the periphery of urban areas nearly five times more than people living in the city center (Börjesson, Eliasson, and Rubensson 2020). The distributional impact therefore depends on where the poor live relative to the rich. Looking at the Buenos Aires transit subsidy program, Bondorevsky (2007) concludes that, although available to all, the program overwhelmingly benefited the middle-income group more than the poor.

Concentrated regional and spatial impacts

Some communities or regions have a heavy specialization in activities with high carbon intensity and will need a place-based approach to prevent the concentrated (and sometimes permanent) impacts that are most unfair and most likely to trigger opposition. In such cases, aggregate impacts become critical, with regions experiencing large increases in unemployment; drops in income, tax revenues, and investments; outmigration of the most skilled workers; and other factors that increase the challenge of transitioning to alternative activities. The experiences of European coal regions, which lost their coal-related revenues and employment decades ago, illustrate how carbon-intensive regions need to adopt an approach that goes beyond individual cases to create new activities and employment opportunities. These insights are relevant not only for other coal regions but also for other areas with concentrated activities, such as the Amazon, where agricultural practices lead to deforestation, or regions or cities with heavy industries, including cement and steel.

European coal transitions can provide lessons for coal regions and areas with concentrated activities in general

Coal transitions can be managed to minimize short-term impacts and prevent long-term effects, but they take time and resources. Even when accelerated, coal transitions typically take decades, with older, poor-performing, or economically unviable mines and power plants closing first. Without proper planning and policies to facilitate the transition, shocks can lead to volatile sociopolitical conditions and, ultimately, economic and social decline in coal regions. The "unmanaged shock" of the British coal transition (figure 4.4) provides examples, in the mining strikes and industrial disputes in response to coal closure plans in the 1970s and 1980s, and the long-term impacts on former coal communities, which continue to suffer from lower job density, worse health outcomes, and higher unemployment, deprivation, and depopulation than the national average (Brauers, Oei, and Walk 2020). The more gradual approach adopted by Germany was expensive—estimated at €38 billion between 2006 and 2018 alone—but has resulted in better socioeconomic outcomes (Oei, Brauers, and Herpich 2019). The Dutch 10-year coal phase-out, which included substantial support for workers who lost their jobs and was supported by the trade union, shows that a well-planned transition does not have to have severe long-term adverse impacts or high costs (World Bank 2021).

Because a smooth transition is easier to manage, starting coal transition planning early and in a participatory and comprehensive way, even before significant negative impacts are visible, can have large benefits. This planning can include early efforts to gradually reduce coal production and consumption to smooth the transition, prevent lock-in effects, reduce stranded asset risks, and favor diversification, thus avoiding industrial concentration. Successful strategies also tend to combine structural reform with more targeted support—for example, adopting structural policies that are geared toward increasing resilience to shocks, by improving access to financial instruments and borrowing; strengthening social safety nets, critical infrastructure and related services, and health care; facilitating greater labor market flexibility and mobility; and creating alternative employment by incentivizing economic innovation and diversification. Targeted policies aimed at affected workers, such as early retirement packages or financial and reemployment support, can exist alongside broader community- or regional-level initiatives, such as skills training, investments in human capital, local economic development

FIGURE 4.4. Coal production and employment in Germany and the United Kingdom, 1958–2018

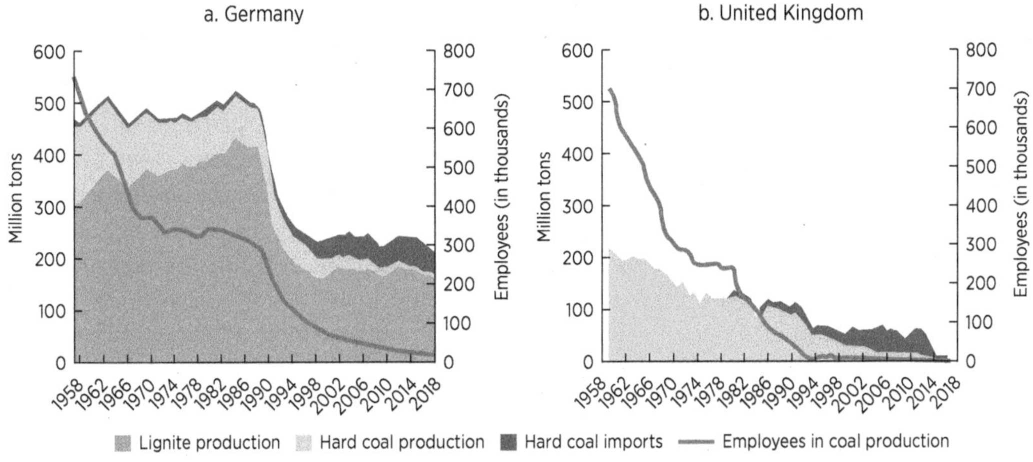

Source: World Bank 2023b.
Note: The United Kingdom phased out coal employment in about 30+ years, whereas it took Germany 60+ years, despite similar economic conditions.

programs, and environmental regeneration. Many developing countries have embarked on their coal transition process and have a unique opportunity to identify highly localized, anticipated economic and social impacts early on, with a view to developing upstream transition policies with the participation of key stakeholders.

Managing major economic transitions must consider effects on labor; social, human, and economic development; local ownership, participation, and mobilization; stakeholder inclusion; and inclusiveness. Rather than adopting simple compensation mechanisms that focus only on employment impacts, successful transitions include targeted social, human, and economic development interventions. These considerations ensure that transition planning can be part of—or a catalyst for—regional and national socioeconomic development plans, including attracting public financing and private investments. Common elements of such transitions include investing in and putting measures in place to improve infrastructure; developing policies and regulations to attract new businesses, education, and skills programs; supporting research and development; and expanding soft location factors in mining regions—such as tertiary education institutions and cultural, leisure, and natural infrastructure—to attract the inward (and prevent the outward) migration of people, business, and investment.

Local ownership of, participation in, and early mobilization for policy design and implementation are important. Local economic development and diversification are key, and policy design needs to respond to local needs and wants. In coal regions, economic structures tend to be concentrated around coal and related industries, and coal tends to have cultural importance. Inclusive processes, local leadership, and mobilizing public, private, nongovernmental, and other actors can help develop locally relevant and responsive transition plans. Building consensus around the need to transition, and developing policies that are guided by community needs and visions of an attractive alternative, contributes to political acceptability. In Germany's Ruhr coal region, for example, transforming previous industrial sites into landmarks or cultural sites made the transition tangible; this initiative marked a break with the past and unveiled a more forward-looking vision for the region but still maintained a distinct local identity. Other initiatives included opening universities, expanding the education system, and improving transportation infrastructure (Arora and Schroeder 2022).

Mainstreaming gender considerations in policy development can produce more inclusive outcomes. When it comes to support for workers, it is worth noting that transitions also affect many indirect jobs. Policies that support only miners may contribute to excluding and disadvantaging the female workers who rely on the coal sector. Evidence also shows that job transitions can contribute to crowding out female workers when competition for limited jobs increases—for example, before Romania restructured its mining sector, women accounted for 16 percent of the workforce; seven years later, this proportion had reduced to just 7 percent (Lahiri-Dutt et al. 2022). Mainstreaming gender considerations in policy development can help ensure gender-inclusive transitions and economic development pathways. Policies that include psychological health and support for household or family members are other gender-sensitive approaches.

Environmental rehabilitation and regeneration are key enablers of alternative economic development. Coal mining, power generation, and related industries cause significant environmental degradation that can limit the potential for alternative economic activities, such as farming or tourism. Communities living in coal regions can experience long-term impacts of such environmental pollution—including negative health impacts, poor water quality, soil contamination, and safety issues—long after the industry has left.

Historical experiences suggest that governments do not always plan adequately for this element of coal (or other industrial) transitions, hindering the longer-term economic renewal of former coal-dependent regions. Along with strengthening regulations and enforcement mechanisms, including clearly establishing "polluter pays" mechanisms in mining licenses to create appropriate incentives, early and progressive rehabilitation efforts can ensure that coal mining and other companies address environmental legacies before they leave the area.

Nationally coordinated fiscal support plays an important role. Despite the importance of bottom-up and locally led approaches, a successful coal transition requires significant national-level support and coordination. Sectoral adjustments or transitions require considerable resources: in most historical cases (primarily from the European Union), national governments have had to cover them, because of insufficient local resources and local fiscal capacity eroded by the economic impact of the coal transition. A full cost-benefit analysis needs to include the cost of direct and hidden coal sector subsidies, and the costs of environmental and public health externalities from coal mining and combustion, which are typically several magnitudes larger than the direct economic benefits of coal use or the fiscal costs of the coal transition. Countries can use carbon pricing, levies, taxes, and other tools to compensate for these costs and raise revenues to pay for the coal transition. National coordination is especially important in this regard, to ensure that raised funds are appropriately directed and fully used.

Place-based policies can help balance spatial and regional policy effects

Outmigration can help people adjust to shocks, but not without costs or limits. Although many studies highlight the role of labor mobility in adjusting to local shocks (Duranton and Venables 2020; Hornbeck 2012), that role is often limited. Bartik (2020) observes restricted outmigration in depressed areas of the United States, even with subsidies. Grover, Lall, and Maloney (2022) also find that mobility is low in many developing countries, for various reasons, including skill mismatches, a lack of finance and information, a reluctance to sell land at a loss, attachment to places, and explicit restrictions on mobility through laws and regulations. In some cases, outmigration can make the situation worse for the population left behind. For example, Beaudry, Green, and Sand (2018) find that, on average, outmigration in the United States reduces labor demand and supply in a similar way, so it does not reduce unemployment for the local population. Although it is generally preferable to invest in people instead of places (World Bank 2009), governments can justify place-based interventions that reduce barriers to or the costs of migration, increase spatial equity, or help fulfill the economic potential of affected regions.

Widely used to support the transition of distressed communities, place-based policies can include a range of measures, from tax incentives and expenditures to manufacturing extension and training programs. As discussed in Duranton and Venables (2020) and Grover, Lall, and Maloney (2022), these interventions need to include multiple instruments, such as transportation investments to improve connections within lagging regions and between lagging and more prosperous regions; fiscal incentives and direct service provisions; and a package of measures to foster skills, enterprise development, and innovation. To be successful, however, transitions must coordinate actions across these functions, as illustrated by an early pilot for the Integrated Rural Development programs of the 1970s and 1980s in Colombia, where successes in some dimensions (technology assistance and input component) were negated by failure in others (market integration).

Place-based policies can have a permanent impact on employment, but this impact, and the policies' cost-effectiveness, depend on design and scale. Bartik (2020) finds permanent positive impact of job shocks in depressed areas, with each job creation increasing employment by 1.2 jobs and, if new jobs are in the tradable sector, by 1.5–2.5 jobs. Grover, Lall, and Maloney (2022) find that such elasticities are even larger in lower-income countries, exceeding five jobs in Costa Rica, the Dominican Republic, and Mexico (figure 4.5), partly because of the large agriculture labor reserve, magnified in depressed areas by the availability of workers from declining industries.

Considerable evidence demonstrates that tax incentives alone are not enough for a policy to succeed (Duranton and Venables 2018). Looking at data across 77 countries, Farole (2011) finds that infrastructure and trade facilitation have a significant positive impact but that tax and other financial incentives are less important. In Bangladesh, studies suggest that special economic zones deliver benefits by offering well-serviced land to manufacturing investors (Duranton and Venables 2018). In the United States, Bartik (2020) finds that tax incentives cost US$110,000–US$200,000 per job created; however, infrastructure programs like the Tennessee Valley Authority, customized public services to business, and investments that make land available for business development, have much lower costs per job created, at US$77,000, US$35,000, and US$13,000, respectively.

Research on the costs and benefits of place-based policies is inconclusive, suggesting that results depend on scale and design, and that place-based policies need to tackle well-identified market or coordination failures. Bartik (2020) estimates the social benefits of each job created in a depressed area at US$240,000–US$400,000 (with 5 percent

FIGURE 4.5. Job multipliers for creating jobs in the tradable sector

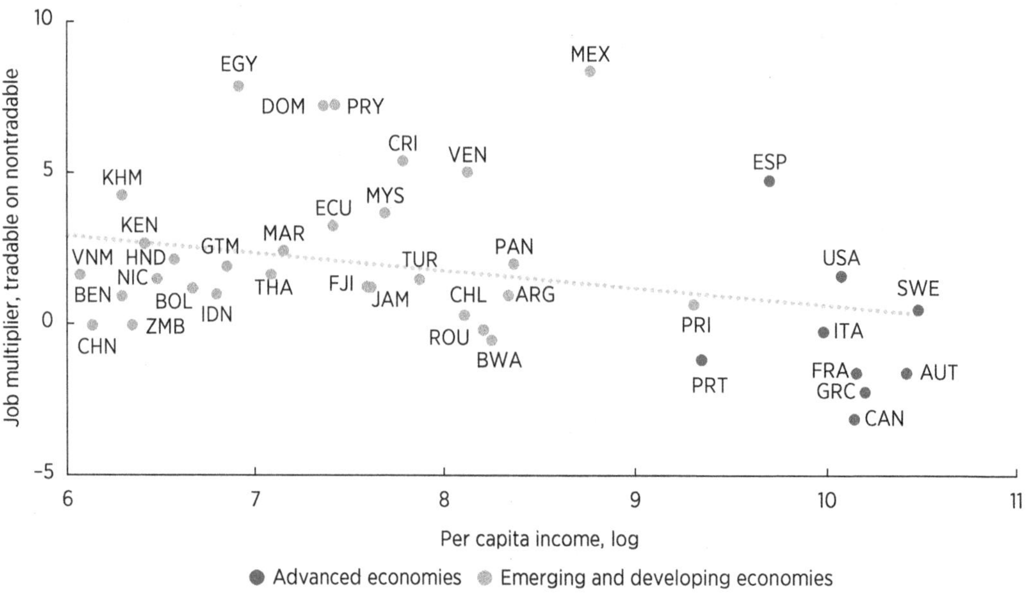

Source: Grover, Lall, and Maloney 2022.
Note: The figure plots how much employment in the local nontradeable sector is generated by a 1 percent increase in employment in the tradable sector. Estimates for the United States are from Moretti 2010. GDP per capita data are from the 2018 World Bank World Development Indicators (accessed in 2021). Estimates use the Integrated Public Use Microdata Series (IPUMS) census data. Data labels use International Organization for Standardization country codes.

or 3 percent discount rates), leading to a benefit-cost ratio of about 1.5 for tax incentives, and higher ratios for infrastructure programs, customized public services, and investments in land availability. Although those estimates are very uncertain and come from a single high-income country, they nevertheless emphasize the potential value of well-designed policies to support job creation in depressed areas; benefits should be larger if the region has a latent comparative advantage—for example, because it is well connected to other economic centers or already has a large labor market. Distributional concerns and political economy considerations would only make these policies more attractive. Grover, Lall, and Maloney 2022 highlight that decisions on whether and how to design and implement place-based policies should be based on identifying the market or coordination failures to be overcome and on a rigorous evaluation of the costs and expected benefits of the interventions. As noted in World Bank Group (2018), outmigration is likely to play an important role in some areas, especially for coal communities, which face the most severe challenges to reinventing themselves. The potential to create jobs in these areas can be limited by a narrow economic base, geographic isolation, wage differences between coal mining and alternative professions, and vocational identity.

References

Aguiar, A., M. Chepeliev, E. L. Corong, R. McDougall, and D. van der Mensbrugghe. 2019. "The GTAP Data Base: Version 10." *Journal of Global Economic Analysis* 4 (1): 1–27. https://doi.org/10.21642/JGEA.040101AF.

Alternburg, T., and C. Assmann. 2017. "Green Industrial Policy." *Concept, Policies, Country Experiences*. Geneva and Bonn: UN Environment.

Andersson, J., and G. Atkinson. 2020. "The Distributional Effects of a Carbon Tax: The Role of Income Inequality." Working Paper 349, Grantham Research Institute on Climate Change and the Environment, London. https://www.lse.ac.uk/granthaminstitute/publication/the-distributional-effects-of-a-carbon-tax-the-role-of-income-inequality/.

Anthoff, D., and R. Hahn. 2010. "Government Failure and Market Failure: On the Inefficiency of Environmental and Energy Policy." *Oxford Review of Economic Policy* 26 (2), 197–224.

Arora, A., and H. Schroeder. 2022. "How to Avoid Unjust Energy Transitions: Insights from the Ruhr Region." *Energy, Sustainability and Society* 12 (1): 19. https://doi.org/10.1186/s13705-022-00345-5.

Ashby, J., A. Braun, T. García, M. del Pilar Guerrero, L. F. Hernandez, C. A. Quirós, and J. I. Roa. 2000. *Investing in Farmers as Researchers. Experience with Local Agricultural Research Committees in Latin America*. CIAT.

Aslund, O., J. Osth, and Y. Zenou. 2010. "How Important Is Access to Jobs? Old Question—Improved Answer." *Journal of Economic Geography* 10 (3): 389–422. https://doi.org/10.1093/jeg/lbp040.

Azevedo, D., H. Wolff, and A. Yamazaki. 2018. "Do Carbon Taxes Kill Jobs? Firm-Level Evidence from British Columbia." Clean Economy Working Paper Series 18.08, Smart Prosperity Institute.

Baldock, D., and A. Buckwell. 2022. *Just Transition in the EU Agriculture and Land Use Sector.* Institute for European Environmental Policy. https://ieep.eu/publications/just-transition-in-the-eu-agriculture-and-land-use-sector/.

Ballon, P., and J. Cuesta. Forthcoming. "Measuring Social Sustainability: A Multidimensional Approach." Social Sustainability and Inclusion Global Practice, Global Unit, World Bank, Washington, DC.

Banerjee, A., P. Niehaus, and T. Suri. 2019. "Universal Basic Income in the Developing World." *Annual Review of Economics* 11 (1): 959–83. https://doi.org/10.1146/annurev-economics-080218-030229.

Bartik, T. J. 2020. "Using Place-Based Jobs Policies to Help Distressed Communities." *Journal of Economic Perspectives* 34 (3): 99–127. https://doi.org/10.1257/jep.34.3.99.

Beaudry, P., D. A. Green, and B. M. Sand. 2018. "In Search of Labor Demand." *American Economic Review* 108 (9): 2714–57. https://doi.org/10.1257/aer.20141374.

Beck, M., N. Rivers, R. Wigle, and H. Yonezawa. 2015. "Carbon Tax and Revenue Recycling: Impacts on Households in British Columbia." *Resource and Energy Economics* 41: 40–69. https://doi.org/10.1016/j.reseneeco.2015.04.005.

Berryman, A. K., J. Bucker, F. Senra de Moura, P. Mealy, M. del Rio-Chanona, P. Barbrook-Johnson, M. Hanusch, and J. D. Farmer. Forthcoming. "Modelling Labor Market Transitions: The Case of Productivity Shifts in Brazil." World Bank Background Note, *North East Economic Memorandum*.

Beuchelt, T. D., and L. Badstue. 2013. "Gender, Nutrition- and Climate-Smart Food Production: Opportunities and Trade-Offs." *Food Security* 5: 709–21. https://doi.org/10.1007/s12571-013-0290-8.

Bondorevsky, D. 2007. "Un Análisis Distributivo sobre el Efecto de los Subsidios al Transporte Público de Pasajeros entre 2002 y 2006 en la Región Metropolitana de Buenos Aires [A Distributive Analysis of Public Urban Transport Services in the Metropolitan Area of Buenos Aires between 2002 and 2006]." Paper prepared for the World Bank.

Börjesson, M., J. Eliasson, and I. Rubensson, I. 2020. "Distributional Effects of Public Transport Subsidies." *Journal of Transport Geography* 84 (102674). https://doi.org/10.1016/j.jtrangeo.2020.102674.

Botta, E. 2019. *A Review of "Transition Management" Strategies: Lessons for Advancing the Green Low-Carbon Transition.* Paris: OECD Publishing. https://doi.org/10.1787/4617a02b-en.

Boyce, J. K. 2018. "Carbon Pricing: Effectiveness and Equity." *Ecological Economics* 150: 52–61. https://doi.org/10.1016/j.ecolecon.2018.03.030.

Boyd, W. 2010. "Ways of Seeing Environmental Law: How Deforestation Became an Object of Climate Governance." *Ecology Law Quarterly* 37.

Brauers, H., P.-Y. Oei, and P. Walk. 2020. "Comparing Coal Phase-Out Pathways: The United Kingdom's and Germany's Diverging Transitions." *Environmental Innovation and Societal Transitions* 37: 238–53. https://doi.org/10.1016/j.eist.2020.09.001.

Braun A., and D. Duveskog. 2011. "The Farmer Field School Approach: History, Global Assessment and Success Stories." Background paper for the IFAD Rural Poverty Report.

Brenner, M., M. Riddle, and J. K. Boyce. 2007. "A Chinese Sky Trust? Distributional Impacts of Carbon Charges and Revenue Recycling in China." *Energy Policy* 35 (3): 1771–84. https://doi.org/10.1016/j.enpol.2006.04.016.

Cai, M., N. Cusumano, A. Lorenzoni, and F. Pontoni. 2016. "A Comprehensive Ex-Post Assessment of the Italian RES Policy: Deployment, Jobs, Value Added and Import Leakages." IEFE Working Papers 88, Center for Research on Energy and Environmental Economics and Policy, Universita' Bocconi, Milan. https://ideas.repec.org/p/bcu/iefewp/iefewp88.html.

Cameron, C., S. Pachauri, N. D. Rao, D. McCollum, J. Rogelj, and K. Riahi. 2016. "Policy Trade-Offs Between Climate Mitigation and Clean Cook-Stove Access in South Asia." *Nature Energy* 1: 15010. https://www.nature.com/articles/nenergy201510.

Carattini, S., M. Carvalho, and S. Fankhauser. 2018. "Overcoming Public Resistance to Carbon Taxes." *WIREs Climate Change* 9 (5). https://doi.org/10.1002/wcc.531.

Castellanos, K. A., and G. Heutel. 2019. "Unemployment, Labor Mobility, and Climate Policy." NBER Working Paper 25797, National Bureau of Economic Research, Cambridge, MA. http://www.nber.org/papers/w25797.

Coady, D., and N. P. Le. 2020. "Designing Fiscal Redistribution. The Role of Universal and Targeted Transfers." IMF Working Paper 105, International Monetary Fund, Washington, DC. https://doi.org/10.5089/9781513547046.001.

Cormack, Z., and A. Kurewa. 2018. "The Changing Value of Land in Northern Kenya: The Case of Lake Turkana Wind Power." *Critical African Studies* 10 (1): 89–107. https://doi.org/10.1080/21681392.2018.1470017.

Cuesta, J., B. López-Noval, and M. Niño-Zarazúa. 2022. "Social Exclusion. Concepts, Measurement and a Global Estimate." Policy Research Working Paper 10097, World Bank, Washington, DC. http://documents.worldbank.org/curated/en/099935306222234310/IDU095f1e5c6060430499b08d1d05f99fe03c118.

Damania, R., E. Balseca, C. de Fontaubert, J. Gill, K. Kim, J. Rentschler, J. Russ, and E. Zaveri. 2023. *Detox Development: Repurposing Environmentally Harmful Subsidies*. Washington, DC: World Bank. http://hdl.handle.net/10986/39423.

da Silva Freitas, L., L. C. de Santana Ribeiro, K. B. de Souza, and G. J. D. Hewings. 2016. "The Distributional Effects of Emissions Taxation in Brazil and Their Implications for Climate Policy." *Energy Economics* 59: 37–44. https://doi.org/10.1016/j.eneco.2016.07.021.

Datta, A. 2010. "The Incidence of Fuel Taxation in India." *Energy Economics* 32: 26–33. https://doi.org/10.1016/j.eneco.2009.10.007.

del Río, P., and M. Burguillo. 2009. "An Empirical Analysis of the Impact of Renewable Energy Deployment on Local Sustainability." *Renewable and Sustainable Energy Reviews* 13 (6): 1314–25. https://doi.org/10.1016/j.rser.2008.08.001.

Devarajan, S., D. S. Go, S. Robinson, and K. Thierfelder. 2011. "Tax Policy to Reduce Carbon Emissions in a Distorted Economy: Illustrations from a South Africa CGE Model." *B.E. Journal of Economic Analysis and Policy* 11 (1). https://doi.org/10.2202/1935-1682.2376.

Dissou, Y., and M. S. Siddiqui. 2014. "Can Carbon Taxes Be Progressive?" *Energy Economics* 42: 88–100. https://doi.org/10.1016/j.eneco.2013.11.010.

Dorband, I. I., M. Jakob, M. Kalkuhl, and J. C. Steckel. 2019. "Poverty and Distributional Effects of Carbon Pricing in Low- and Middle-Income Countries—A Global Comparative Analysis." *World Development* 115: 246–57. https://doi.org/10.1016/j.worlddev.2018.11.015.

Dorband, I. I., M. Jakob, J. C. Steckel, and H. Ward. 2022. "Double Progressivity of Infrastructure Financing through Carbon Pricing—Insights from Nigeria." *World Development Sustainability* 1: 100011. https://doi.org/10.1016/j.wds.2022.100011.

Dorband, I. I. Forthcoming. "Distributional Effects of Climate and Development Policies across Households and Workers in Low- and Middle-Income Countries."

Dorin, B., J.-C. Hourcade, and M. Benoit-Cattin. 2013. "A World without Farmers? The Lewis Path Revisited." CIRED Working Paper No. 47-2013, Centre International de Recherches sur l'Environnement et le Développement, Nogent-sur-Marne, France. https://agritrop.cirad.fr/570557/1/document_570557.pdf.

Douenne, T. 2020. "The Vertical and Horizontal Distributive Effects of Energy Taxes: A Case Study of a French Policy." *Energy Journal* 41 (3). https://doi.org/10.5547/01956574.41.3.tdou.

Duranton, G., and A. J. Venables. 2018. "Place-Based Policies for Development." NBER Working Paper 24562, National Bureau of Economic Research, Cambridge, MA.

Duranton, G., and A. J. Venables. 2020. "Place-Based Policies fof Development." In *Handbook of Regional Science*, edited by M. Fisher and P. Nijkamp. Berlin: Springer-Verlag.

Dussaux, D. 2020. "The Joint Effects of Energy Prices and Carbon Taxes on Environmental and Economic Performance: Evidence from the French Sector." OECD Environment Working Paper 154, Organisation for Economic Co-operation and Development, Paris. https://doi.org/10.1787/b84b1b7d-en.

Edwards, P. E. T., A. E. Sutton-Grier, and G. E. Coyle. 2013. "Investing in Nature: Restoring Coastal Habitat Blue Infrastructure and Green Job Creation." *Marine Policy* 38: 65–71. https://doi.org/10.1016/j.marpol.2012.05.020.

Erman, A., S. A. De Vries Robbe, S. Thies, K. Kabir, and M. Maruo. 2021. "Gender Dimensions of Disaster Risk and Resilience. Existing Evidence." World Bank, Washington, DC. http://hdl.handle.net/10986/35202.

Fagbemi, F., T. T. Osinubi, and O. A. Adeosun. 2022. "Enhancing Sustainable Infrastructure Development: A Boon to Poverty Reduction in Nigeria." *World Development Sustainability* 1: 100006. https://doi.org/10.1016/j.wds.2022.100006.

Farole, T. 2011. *Special Economic Zones in Africa: Comparing Performance and Learning from Global Experience*, Directions in Development. Washington, DC: World Bank. http://hdl.handle.net/10986/2268.

Feindt, S., U. Kornek, J. M. Labeaga, T. Sterner, and H. Ward. 2021. "Understanding Regressivity: Challenges and Opportunities of European Carbon Pricing." *Energy Economics* 103: 105550. https://doi.org/10.1016/j.eneco.2021.105550.

Feng, K., K. Hubacek, D. Guan, M. Contestabile, J. Minx, and J. Barrett, J. 2010. "Distributional Effects of Climate Change Taxation: The Case of the UK." *Environmental Science & Technology* 44 (10): 3670–76. https://doi.org/10.1021/es902974g.

Feng, K., K. Hubacek, Y. Liu, E. Marchán, and A. Vogt-Schilb. 2018. "Managing the Distributional Effects of Energy Taxes and Subsidy Removal in Latin America and the Caribbean." *Applied Energy* 225: 424–36. https://doi.org/10.1016/j.apenergy.2018.04.116.

Franklin, S. 2018. "Location, Search Costs and Youth Unemployment: Experimental Evidence from Transport Subsidies." *Economic Journal* 128 (614): 2353–79. https://doi.org/10.1111/ecoj.12509.

Friis-Hansen E., and D. Duveskog. 2012. "The Empowerment Route to Well-being: An Analysis of Farmer Field Schools in East Africa." *World Development* 40: 414–27.

Fuglie, K., M. Gautam, A. Goyal, and W. F. Maloney. 2020. *Harvesting Prosperity: Technology and Productivity Growth in Agriculture*. World Bank, Washington, DC. http://hdl.handle.net/10986/32350.

Fuglie, K., S. Ray, U. L. C. Baldos, and T. W. Hertel. 2022. "The R&D Cost of Climate Mitigation in Agriculture." *Applied Economic Perspectives and Policy* 44 (4): 1955–74. https://doi.org/10.1002/aepp.13245.

Gautam, M., D. Laborde, A. Mamun, W. Martin, V. Pineiro, and R. Vos. 2022. "Repurposing Agricultural Policies and Support: Options to Transform Agriculture and Food Systems to Better Serve the Health of People, Economies and the Planet." World Bank, Washington, DC. http://hdl.handle.net/10986/36875.

Godinho, C. 2022. "What Do We Know about the Employment Impacts of Climate Policies? A Review of the Ex Post Literature." *WIREs Climate Change* 13 (6). https://doi.org/10.1002/wcc.794.

Goulder, L. H., M. A. C. Hafstead, G. Kim, and X. Long. 2019. "Impacts of a Carbon Tax across US Household Income Groups: What Are the Equity-Efficiency Trade-Offs?" *Journal of Public Economics* 175: 44–64. https://doi.org/10.1016/j.jpubeco.2019.04.002.

Goulder, L. H., M. A. C. Hafstead, and R. C. Williams. 2016. "General Equilibrium Impacts of a Federal Clean Energy Standard." *American Economic Journal: Economic Policy* 8 (2): 186–218. https://doi.org/10.1257/pol.20140011.

Grainger, C. A., and C. D. Kolstad. 2010. "Who Pays a Price on Carbon?" *Environmental and Resource Economics* 46: 359–76. https://doi.org/10.1007/s10640-010-9345-x.

Greve, H., and J. Lay. 2023. "'Stepping Down the Ladder': The Impacts of Fossil Fuel Subsidy Removal in a Developing Country." *Journal of the Association of Environmental and Resource Economists* 10 (1): 121–58. https://doi.org/10.1086/721375.

Grover, A. G., S. V. Lall, and W. F. Maloney. 2022. *Place, Productivity and Prosperity: Revisiting Spatially Targeted Policies for Regional Development—Overview (Vol. 2)*. Washington, DC: World Bank. http://documents.worldbank.org/curated/en/099130001182297916/P1725410e343500aa0a3de030d5f1599b46.

Gusdorf, F., and S. Hallegatte. 2007. "Compact or Spread-Out Cities: Urban Planning, Taxation and the Vulnerability to Transportation Shocks." *Energy Policy* 35 (10): 4826–38. https://doi.org/10.1016/j.enpol.2007.04.017.

Hallegatte, S., M. Fay, and A. Vogt-Schilb. 2013. "Green Industrial Policies: When and How." Policy Research Working Paper 6677, World Bank, Washington, DC.

Hanna, R., and B. A. Olken. 2018. "Universal Basic Incomes versus Targeted Transfers: Anti-Poverty Programs in Developing Countries." *Journal of Economic Perspectives* 32 (4): 201–26. https://doi.org/10.1257/jep.32.4.201.

Hassan, M., and W. Prichard. 2016. "The Political Economy of Domestic Tax Reform in Bangladesh: Political Settlements, Informal Institutions and the Negotiation of Reform." *Journal of Development Studies* 52 (12): 1704–21. https://doi.org/10.1080/00220388.2016.1153072.

Helm, D. 2010. "Government Failure, Rent-Seeking, and Capture: The Design of Climate Change Policy." *Oxford Review of Economic Policy* 26 (2): 182–96.

Hille, E., and P. Möbius. 2019. "Do Energy Prices Affect Employment? Decomposed International Evidence." *Journal of Environmental Economics and Management* 96 (1–21). https://doi.org/10.1016/j.jeem.2019.04.002.

Hornbeck, R. 2012. "The Enduring Impact of the American Dust Bowl: Short- and Long-Run Adjustments to Environmental Catastrophe." *American Economic Review* 102 (4):1477–507.

Horowitz, J. 2022. "Ditching Fossil Fuel Subsidies Can Trigger Unrest. Keeping Them Will Kill the Climate." *CNN Business,* January 20, 2022. https://edition.cnn.com/2022/01/20/energy/oil-subsidies-unrest-climate/index.html.

Huesca-Pérez, M. E., C. Sheinbaum-Pardo, and J. Köppel. 2016. "Social Implications of Siting Wind Energy in a Disadvantaged Region—The Case of the Isthmus of Tehuantepec, Mexico." *Renewable and Sustainable Energy Reviews* 58: 952–65. https://doi.org/10.1016/j.rser.2015.12.310.

Humphries S., L. Classen, J. Jiménez, F. Sierra, O. Gallardo, and M. Gómez. 2012. "Opening Cracks for the Transgression of Social Boundaries: An Evaluation of the Gender Impacts of Farmer Research Teams in Honduras." *World Development* 40: 2078–95.

IFC (International Finance Corporation). 2019. *The Dirty Footprint of the Broken Grid. The Impacts of Fossil Fuel Back-up Generators in Developing Countries.* Washington, DC: International Finance Corporation. https://www.ifc.org/wps/wcm/connect/Industry_EXT_Content/IFC_External_Corporate_Site/Financial+Institutions/Resources/Dirty-footprint-of-broken-grid.

Ikeda, K. 2009. "How Women's Concerns Are Shaped in Community-Based Disaster Risk Management in Bangladesh." *Contemporary South Asia* 17 (1): 65–78. https://doi.org/10.1080/09584930802624679.

Javaid, A., F. Creutzig, and S. Bamberg. 2020. "Determinants of Low-Carbon Transport Mode Adoption: Systematic Review of Reviews." *Environmental Research Letters* 15 (10): 103002. https://doi.org/10.1088/1748-9326/aba032.

Jin, J., and K. Paulsen. 2017. "Does Accessibility Matter? Understanding the Effect of Job Accessibility on Labour Market Outcomes." *Urban Studies* 55 (1). https://doi.org/10.1177/0042098016684099.

Johnson, O., T. Altenburg, and H. Schmitz. 2014. "Rent Management Capabilities for the Green Transformation." In *Green Industrial Policy in Emerging Countries,* edited by A. Pegels. Abingdon: Routledge.

Juhász, R., N. J. Lane, and D. Rodrik. 2023. "The New Economics of Industrial Policy." NBER Working Paper 31538, National Bureau of Economic Research, Cambridge, MA.

Jumani, S., S. Rao, S. Machado, and A. Prakash. 2017. "Big Concerns with Small Projects: Evaluating the Socio-Ecological Impacts of Small Hydropower Projects in India." *Ambio* 46 (4): 500–11. https://doi.org/10.1007/s13280-016-0855-9.

Kabir, K., S. A. De Vries Robbe, and C. Godinho. Forthcoming. "Climate Mitigation Policies in Agriculture: A Review of Socio-Political Barriers."

Kerkhof, A. C., H. C. Moll, E. Drissen, and H. C. Wilting. 2008. "Taxation of Multiple Greenhouse Gases and the Effects on Income Distribution: A Case Study of the Netherlands." *Ecological Economics* 67 (2): 318–26. https://doi.org/10.1016/j.ecolecon.2007.12.015.

Kerkhof, A. C., S. Nonhebel, and H. C. Moll. 2009. "Relating the Environmental Impact of Consumption to Household Expenditures: An Input–Output Analysis." *Ecological Economics* 68 (4): 1160–70. https://doi.org/10.1016/j.ecolecon.2008.08.004.

Klenert, D., L. Mattauch, E. Combet, O. Edenhofer, C. Hepburn, R. Rafaty, and N. Stern. 2018. "Making Carbon Pricing Work for Citizens." *Nature Climate Change* 8 (8): 669–77. https://doi.org/10.1038/s41558-018-0201-2.

Köppl, A., and M. Schratzenstaller. 2022. "Carbon Taxation: A Review of the Empirical Literature." *Journal of Economic Surveys* 37 (4): 1353–88. https://doi.org/10.1111/joes.12531.

Lahiri-Dutt, S., S. Dowling, D. Pasaribu, A. Chowdhury, and R. Talukdar. 2022. *Just Transition for All: A Feminist Approach for the Coal Sector.* Washington, DC: World Bank. http://documents.worldbank.org/curated/en/099405206192237419/P1711940b3d5590820b3480a4662ace12ea.

Lakhanpal, S. 2019. "Contesting Renewable Energy in the Global South: A Case-Study of Local Opposition to a Wind Power Project in the Western Ghats of India." *Environmental Development* 30: 51–60. https://doi.org/10.1016/j.envdev.2019.02.002.

Larson, A. M., T. Dokken, A. E. Duchelle, S. Atmadja, I. A. P. Resosudarmo, P. Cronkleton, M. Cromberg, W. Sunderlin, A. Awono, and G. Selaya. 2015. "The Role of Women in Early REDD+ Implementation: Lessons for Future Engagement." *International Forestry Review* 17 (1): 43–65.

Leistritz, F. L., and R. C. Coon. 2009. "Socioeconomic Impacts of Developing Wind Energy in the Great Plains." *Great Plains Research: A Journal of Natural and Social Sciences* 997. https://core.ac.uk/reader/17249390.

Lindberg, M. B., and L. Kammermann. 2021. "Advocacy Coalitions in the Acceleration Phase of the European Energy Transition." *Environmental Innovation and Societal Transitions* 40: 262–82. https://doi.org/10.1016/j.eist.2021.07.006.

Liotta, C., P. Avner, V. Viguié, H. Selod, and S. Hallegatte. 2022. "Climate Policy and Inequality in Urban Areas: Beyond Incomes." Policy Research Working Paper 10185, World Bank, Washington, DC. https://doi.org/DOI:10.1596/1813-9450-10185.

Malakar, Y., C. Greig, and E. van de Fliert. 2018. "Resistance in Rejecting Solid Fuels: Beyond Availability and Adoption in the Structural Dominations of Cooking Practices in Rural India." *Energy Research and Social Science* 46: 225–35. https://doi.org/10.1016/j.erss.2018.07.025.

Marin, G., and F. Vona. 2019. "Climate Policies and Skill-Biased Employment Dynamics: Evidence from EU Countries." *Journal of Environmental Economics and Management* 98: 102253. https://doi.org/10.1016/j.jeem.2019.102253.

Markandya, A., I. Arto, M. González-Eguino, and M. V. Román. 2016. "Towards a Green Energy Economy? Tracking the Employment Effects of Low-Carbon Technologies in the European Union." *Applied Energy* 179: 1342–50. https://doi.org/10.1016/j.apenergy.2016.02.122.

Metcalf, G. E. 2008. "Designing a Carbon Tax to Reduce U.S. Greenhouse Gas Emissions." *Review of Environmental Economics and Policy* 3 (1): 63–83. https://doi.org/10.1093/reep/ren015.

Metcalf, G. E., and J. H. Stock. 2020. "Measuring the Macroeconomic Impact of Carbon Taxes." *AEA Papers and Proceedings* 110: 101–106. https://doi.org/10.1257/pandp.20201081.

Missbach, L., J. C. Steckel, and A. Vogt-Schilb. 2022. "Cash Transfers in the Context of Carbon Pricing Reforms in Latin America and the Caribbean." IDB Working Paper Series, Inter-American Development Bank, Washington, DC. http://dx.doi.org/10.18235/0004568.

Moretti, E. 2010. "Local Multipliers." *American Economic Review* 100 (2): 373–77.

Muller, C., and H. Yan. 2018. "Household Fuel Use in Developing Countries: Review of Theory and Evidence." *Energy Economics* 70: 429–39. https://doi.org/10.1016/j.eneco.2018.01.024.

Nell, A., D. Herszenhut, C. Knudsen, S. Nakamura, M. Saraiva, and P. Avner. 2023. "Carbon Pricing and Transit Accessibility to Jobs: Impacts on Inequality in Rio de Janeiro and Kinshasa." Policy Research Working Paper 10341, World Bank, Washington, DC. https://doi.org/10.1596/1813-9450-10341.

Neven, D. J., and L. H. Röller. 2000. *The Political Economy of Industrial Policy in Europe and the Member States.* Berlin: Edition Sigma.

Nurdianto, D. A., and B. P. Resosudarmo. 2016. "The Economy-Wide Impact of a Uniform Carbon Tax in ASEAN." *Journal of Southeast Asian Economies* 33 (1): 1–22. https://doi.org/10.1353/ase.2016.0009.

Obour, P. B., K. Owusu, E. A. Agyeman, A. Ahenkan, and A. N. Madrid. 2016. "The Impacts of Dams on Local Livelihoods: A Study of the Bui Hydroelectric Project in Ghana." *International Journal of Water Resources Development* 32 (2): 286–300. https://doi.org/10.1080/07900627.2015.1022892.

Oei, P.-Y., H. Brauers, and P. Herpich. 2019. "Lessons from Germany's Hard Coal Mining Phase-Out: Policies and Transition from 1950 to 2018." *Climate Policy* 20. https://doi.org/10.1080/14693062.2019.1688636.

Ohlendorf, N., M. Jakob, J. C. Minx, C. Schröder, and J. C. Steckel. 2021. "Distributional Impacts of Carbon Pricing: A Meta-Analysis." *Environmental and Resource Economics* 78 (1): 1–42. https://doi.org/10.1007/s10640-020-00521-1.

Ortega, M., P. del Río, P. Ruiz, and C. Thiel. 2015. "Employment Effects of Renewable Electricity Deployment. A Novel Methodology." *Energy* 91: 940–51. https://doi.org/10.1016/j.energy.2015.08.061.

Pegels, A., ed. 2014. *Green Industrial Policy in Emerging Countries.* Routledge.

Rao, N. D. 2015. "Climate Change and Poverty: Energy Price Channels." Presentation to the World Bank "Climate Change and Poverty Conference," Washington, DC, February 9. https://www.worldbank.org/content/dam/Worldbank/document/Climate/Climate%20and%20Poverty%20Conference/D1S2_NRao-Poverty%20vs%20Mitigation%20-WB.pdf.

Renner, S. 2018. "Poverty and Distributional Effects of a Carbon Tax in Mexico." *Energy Policy* 112: 98–110. https://doi.org/10.1016/j.enpol.2017.10.011.

Rodrik, D. 2014. "When Ideas Trump Interests: Preferences, Worldviews and Policy Innovations." *Journal of Economic Perspectives* 28 (1): 189–208. https://doi.org/10.1257/jep.28.1.189.

Rozenberg, J., A. Vogt-Schilb, and S. Hallegatte. 2020. "Instrument Choice and Stranded Assets in the Transition to Clean Capital." *Journal of Environmental Economics and Management* 100 (2020): 102183.

Saelim, S. 2019. "Carbon Tax Incidence on Household Consumption: Heterogeneity Across Socio-Economic Factors in Thailand." *Economic Analysis and Policy* 62: 159–74. https://doi.org/10.1016/j.eap.2019.02.003.

Saussay, A., M. Sato, F. Vona, and L. O'Kane. 2022. "Who's Fit for the Low-Carbon Transition? Emerging Skills and Wage Gaps in Job Ad Data." Policy Working Paper, Grantham Centre for Climate Change Economics and the Environment, London School of Economics and Political Science.

Schaffitzel, F., M. Jakob, R. Soria, A. Vogt-Schilb, and H. Ward. 2020. "Can Government Transfers Make Energy Subsidy Reform Socially Acceptable? A Case Study on Ecuador." *Energy Policy* 137: 111120. https://doi.org/10.1016/j.enpol.2019.111120.

Shah, A., and B. Larsen. 1992. "Carbon Taxes, the Greenhouse Effect, and Developing Countries." Policy Research Working Paper WPS957, World Bank, Washington, DC. http://documents.worldbank.org/curated/en/460851468739298164/Carbon-taxes-the-greenhouse-effect-and-developing-countries.

Somanathan E., T. Sterner, T. Sugiyama, D. Chimanikire, N. K. Dubash, J. Essandoh-Yeddu, S. Fifita, L. Goulder, A. Jaffe, X. Labandeira, S. Managi, C. Mitchell, J. P. Montero, F. Teng, and T. Zylicz. 2014: "National and Sub-national Policies and Institutions." In *Climate Change 2014: Mitigation of Climate Change. Contribution of Working Group III to the Fifth Assessment Report of the Intergovernmental Panel on Climate Change* edited by O. Edenhofer, R. Pichs-Madruga, Y. Sokona, E. Farahani, S. Kadner, K. Seyboth, A. Adler, I. Baum, S. Brunner, P. Eickemeier, B. Kriemann, J. Savolainen, S. Schlömer, C. von Stechow, T. Zwickel, and J. C. Minx. Cambridge University Press.

Steckel, J. C., I. I. Dorband, L. Montrone, H. Ward, L. Missbach, F. Hafner, M. Jakob, and S. Renner. 2021. "Distributional Impacts of Carbon Pricing in Developing Asia." *Nature Sustainability* 4 (11): 1005–14. https://doi.org/10.1038/s41893-021-00758-8.

Steckel, J. C., S. Renner, and L. Missbach. 2021. "Distributional Impacts of Carbon Pricing in Low- and Middle-Income Countries." *CESifo Forum, Ifo Institut - Leibniz-Institut Für Wirtschaftsforschung an Der Universität München* 22 (5): 26–32. http://hdl.handle.net/10419/250938.

Terrapon-Pfaff, J., T. Fink, P. Viebahn, and E. M. Jamea. 2019. "Social Impacts of Large-Scale Solar Thermal Power Plants: Assessment Results for the $NOOR_O$ I Power Plant in Morocco." *Renewable and Sustainable Energy Reviews* 113: 109259. https://doi.org/10.1016/j.rser.2019.109259.

Townsend, R., R. M. Benefica, A. Prasann, and M. Lee. 2017. "Future of Food: Shaping the Food System to Deliver Jobs." Working Paper, World Bank, Washington, DC. http://documents.worldbank.org/curated/en/406511492528621198/Future-of-food-shaping-the-food-system-to-deliver-jobs.

van Heerden, J., R. Gerlagh, J. Blignaut, M. Horridge, S. Hess, R. E. E. Mabugu, and M. M. Chitiga. 2005. "Fighting CO2 Pollution and Poverty while Promoting Growth: Searching for Triple Dividends in South Africa." PREM Working Paper No. 05-02. https://doi.org/10.2139/ssrn.849245.

Vogt-Schilb, A., B. Walsh, B. K. Feng, L. Di Capua, Y. Liu, D. Zuluaga, M. Robles, and K. Hubaceck. 2019. "Cash Transfers for Pro-Poor Carbon Taxes in Latin America and the Caribbean." *Nature Sustainability* 2 (10): 941–48. https://doi.org/10.1038/s41893-019-0385-0.

Vona, F. 2019. "Job Losses and Political Acceptability of Climate Policies: Why the 'Job-Killing' Argument Is So Persistent and How to Overturn It." *Climate Policy* 19 (4): 524–32. https://doi.org/10.1080/14693062.2018.1532871.

Wier, M., K. Birr-Pedersen, H. Klinge Jacobsen, and J. Klok. 2005. "Are CO2 Taxes Regressive? Evidence from the Danish Experience." *Ecological Economics* 52 (2): 239–51. https://doi.org/10.1016/j.ecolecon.2004.08.005.

World Bank. 2009. *World Development Report 2009: Reshaping Economic Geography.* Washington, DC: World Bank. https://openknowledge.worldbank.org/entities/publication/58557d74-baf0-5f97-a255-00482909810a.

World Bank. 2011. "Gender Equality and Women's Empowerment in Disaster Recovery." GFDRR Disaster Recovery Guidance Series, World Bank, Washington, DC. https://www.gfdrr.org/en/publication/gender-equality-and-womens-empowerment-disaster-recovery.

World Bank. 2021. *Supporting Transition in Coal Regions: A Compendium of the World Bank's Experience and Guidance for Preparing and Managing Future Transitions.* Washington, DC: World Bank. http://hdl.handle.net/10986/35323.

World Bank. 2023a. "Moving the Needle on Clean Cooking for All." Results brief, January 19, 2023. https://www.worldbank.org/en/results/2023/01/19/moving-the-needle-on-clean-cooking-for-all.

World Bank. 2023b. *Reality Check: Lessons from 25 Policies Advancing a Low-Carbon Future.* Washington, DC: World Bank.

World Bank. Forthcoming. "Climate Change Vulnerability Profiles and Local Adaptation Strategies in Indonesia." World Bank.

World Bank Group. 2018. *Managing Coal Mine Closure. Achieving a Just Transition for All.* Washington, DC: World Bank. http://hdl.handle.net/10986/31020.

World Bank Group. 2022a. *Climate and Development: An Agenda for Action—Emerging Insights from World Bank Group 2021–22 Country Climate and Development Reports.* Washington, DC: World Bank. http://hdl.handle.net/10986/38220.

World Bank Group. 2022b. "China. Country Climate and Development Report. CCDR Series." World Bank, Washington, DC. http://hdl.handle.net/10986/38136.

WRI (World Resources Institute). 2019. *Creating a Sustainable Food Future.* Washington, DC: WRI. https://www.wri.org/research/creating-sustainable-food-future.

Yamazaki, A. 2017. "Jobs and Climate Policy: Evidence from British Columbia's Revenue-Neutral Carbon Tax." *Journal of Environmental Economics and Management* 83: 197–216. https://doi.org/10.1016/j.jeem.2017.03.003.

Yusuf, A. A. 2008. "The Distributional Impact of Environmental Policy: The Case of Carbon Tax and Energy Pricing Reform in Indonesia. Research Report." EEPSEA Research Report rr2008101, Environment and Economy Program for Southeast Asia.

Zhang, G., and N. Zhang. 2020. "The Effect of China's Pilot Carbon Emissions Trading Schemes on Poverty Alleviation: A Quasi-Natural Experiment Approach." *Journal of Environmental Management* 271: 110973. https://doi.org/10.1016/j.jenvman.2020.110973.

5 Policy Process

Using Public Engagement and Communication to Improve Policies and Their Legitimacy

KEY INSIGHTS

Support for or opposition to a policy reform depends not only on the policy's design but also on the process that led to its implementation. Analyzing how institutions and actors' interests, ideas, and influence shape policy processes and outcomes can help identify potential sources of support or opposition and allow governments to adjust policy processes accordingly.

Civic engagement can help build legitimacy and develop working compromises and necessary support for urgent action by mediating distributional conflict, differences in preferences and priorities, and unequal power dynamics. But such engagements create trade-offs because they take time and create uncertainties that may delay or reduce reforms and investments.

Communication can be an effective tool in making the implications of a reform accessible and increasing public support. Translating a policy into understandable and relatable messages through communication campaigns can increase public support and acceptance, and therefore the policy's sustainability.

The previous chapter focused on the challenges and opportunities of policy design to manage the distributional effects of climate action—that is, on actors' *interests*. This chapter looks at public engagement and communication strategies to change *ideas* and *influences*, build societal support for policies, and ensure their legitimacy.

Policy processes: Engaging with support and opposition

Climate policies are made and implemented through processes that are shaped by the political economy context and unfolding dynamics between actors. Recommendations on climate policies often assume linear and evidence-based policy processes, bypassing the political realities of decision-making and implementation. But, as shown in the 25 case studies explored in the *Reality Check* companion report (World Bank 2023), policy processes are often shaped by the political economy, and outcomes rarely represent the first-best policy solution. Rather, policies tend to be compromises developed because of their ease of implementation and their acceptability. Unsurprisingly, when policy recommendations emerge from traditional approaches, they do not always find the levels of political support or social acceptability required. Adding in analysis of how institutions and actors' interests, ideas, and influence shape policy processes and outcomes can complement the traditional approach to policy making.

A political economy lens can provide important information on the where, how, who, and why of policy reforms. Political economy analysis tools—such as institutional and stakeholder analysis, power mapping, and opinion surveys—can help policy makers answer key questions about the policy process, such as the following:

- Where and how does policy making take place?
- Who is involved, and how much influence do they have over others or the process?
- What do they know, think, and feel about the policy process, problem, and solutions?
- Which individual or collective interests might be affected by the problem or proposed solutions?
- Why do actors support or oppose climate reforms or policy?

Armed with this information, policy makers can identify potential sources of support or opposition and can adjust policy processes accordingly. Certain features of the policy process can make the difference between success and failure, depending on the extent to which they magnify or mitigate opposition and support. But it is not as simple as tipping the balance in favor of winners and supporters. For a policy decision to be acceptable and implementable, most actors in a society need to consider the process to be credible, fair, and acceptable, especially those actors who stand to lose or who oppose the policy (Barron et al. 2023)—see box 5.1. The social and political acceptability of climate policy is not only about choosing the "right" objectives and instruments while managing effects; it is also about how these decisions are made.

Governments can use features of the policy process to strengthen the acceptability of outcomes. Opposition to climate reforms arises for many reasons but mostly because actors believe that they will be negatively affected, that outcomes will be unfair, or that they have been excluded from the policy process. Low levels of trust in government can magnify these concerns, for instance, if people feel they are not represented or that elected officials are unaccountable. To overcome such challenges, governments will have to mediate and broker compromises between groups, while ensuring that these groups accept the process and outcomes as fair and credible.

Although a necessary part of the policy reform process, public engagement strategies have some limitations in generating reform support. Studies have shown that strong norms, combined with attitudes like NIMBYism,[1] can present challenges in building support for policy reform and undermine the momentum for development. Although feedback and information sessions can help increase participation and inclusion, they often fail to make citizens feel genuinely heard. Such sessions can also fall short on encouraging citizens to understand the competing pressures policy makers and other stakeholders might face, hindering the citizens' willingness to compromise (Doberstein 2020). Because little empirical evidence exists on the effects of public involvement in decision-making, the positive effects expected in theory from such a strategy—such as greater overall satisfaction—do not necessarily occur (Rowe et al. 2008). To meet the demands of democratic, equitable, and inclusive decision-making, policy makers should therefore be

BOX 5.1

Process legitimacy: How decisions are made

One of four critical dimensions of social sustainability—alongside social cohesion, inclusion, and resilience—*process legitimacy* is about how policy making and implementation are done, their consistency with a given context, and their perceived legitimacy. Specifically, it has to do with the extent to which actors in society accept who has authority, the goals they formulate, and how they make and implement decisions. Process legitimacy is strong when actors believe that decisions are made by credible authorities in ways that align with their values and reflect accepted rules and norms relating to decision-making, including around measures that support conflict resolution and compromise. Transparent and participatory processes, and desirable or acceptable outcomes, can enhance process legitimacy. This is especially important when policies incur costs and when inclusion of and engagement with potential policy losers are critical. Five common drivers strengthen process legitimacy:

1. *Credibility of decision-makers.* The power to make decisions, set policies, and implement programs gains legitimacy when it stems from an accepted source, such as an election, formal or informal designation, or technical expertise.
2. *Consistency with agreed-upon rules.* The rules followed to make and implement decisions gain legitimacy when they reflect established methods or approaches that a community or society agrees to be acceptable and credible, such as legal precedents, professional standards, procedural guidelines, informal traditions, or customs.
3. *Consistency with societal values.* Policies gain legitimacy when they respect or follow people's beliefs or moral convictions about what and how things should be done. This includes religious, philosophical, and ideological convictions, and widely respected but nonbinding rules, such as international rules regarding war crime or human rights laws.
4. *Perceived benefits for the affected population.* Policies and programs gain acceptance and legitimacy, even if some people regard them as dubious or morally fraught, as long as key stakeholders are convinced that they are (or will soon be) richer, safer, or better off in some other way. In this sense, the ends justify, or legitimize, the means.
5. *Participation and transparency.* Dialogue, engagement, feedback, and partnership between authority figures and members of a given community or society—coupled with open and transparent decision-making—can generate legitimacy, especially when there are disagreements or tensions.

Source: Barron et al. 2023.

aware of the possible limitations of public engagement strategies and consider a combination of tested approaches to enhance their success.

The climate crisis and related symptoms—such as weather extremes and health impacts—require urgent action, which can create tensions with the time needed for meaningful engagement, especially when many stakeholders are involved. Expanding public transportation, developing renewable energy infrastructure, and other actions to address the impact of climate change can affect the population through different channels. If not addressed properly, these effects can spur protest and opposition. For example, energy security concerns caused by the war in Ukraine gave a huge push to the expansion of renewables in countries formerly dependent on energy imports from the Russian Federation. This expansion of green energy has been facilitated by measures to ease planning and permitting processes, including the streamlining of impact assessments, which some have criticized for shortening and devaluing participatory processes (Geißler and Jiricka-Pürrer 2023; Gonzales and Sobrini 2023). In the United Kingdom, considerations of replacing environmental impact assessments and strategic environmental impact assessments with environmental outcome reports to shorten planning consent procedures have likewise been met with resistance, because of the expectation that the use of such reports will significantly reduce public participation in decision-making (Fischer 2022). Governments' policy responses to the 2020 pandemic, including restrictions on freedom of movement and assembly, were similarly widely criticized for lacking legitimacy (Armeni and Lee 2021). Finding the right balance between enough time for public engagement and the time constraints of developing a green technocratic response presents a trade-off with a still unknown optimum.

Building public support through citizen engagement and strategic communication

Some people claim that government action on climate has been slow because citizens have not demonstrated demand for it and because politicians fear adopting unpopular measures and not bringing voters with them (Wilson 2018). But evidence suggests that most people in most countries are concerned about climate change and support more ambitious climate action. An international survey of 1.2 million people in 50 countries finds that two-thirds believe climate change is a global emergency (UNDP and University of Oxford 2021). In the United States, concern over climate change remains at an all-time high: more than half of all US citizens believe they are being harmed "right now" by climate impacts. These trends are mirrored in China, Sweden, and the United Kingdom, where nearly 50 percent of people polled are "extremely" or "very" worried about climate change and nearly 60 percent want to see urgent action to address it (CAST 2021). In Brazil, concerns over climate change are even higher, at 75 percent (ITS 2021).

Several challenges contribute to a disconnect between this demand and the reality of climate institutions today. First, as discussed in chapter 4, beneficiaries of climate action are dispersed within and between societies—indeed, those who will benefit most are yet to be born—whereas costs can be very concentrated. Second, citizens can be discouraged by the little direct control they have, or think they have, over levers of change, especially when trust in government is low. Finally, the scale of the challenge and its solutions can be overwhelming, yet mainstream and social media often provide little information that could help people make sense of it. This unequal access to information can make it difficult for the public to understand some of the complex technical issues, especially because most individuals or communities lack time and resources. Civic engagement and well-designed policy communication can be used as strategic tools to encourage citizens' participation in decision-making processes, build legitimacy, develop compromises, and increase ownership and public policy support. The following sections introduce underlying concepts and best practices from real-world examples.

Public perceptions: When winners feel like losers

When assessing the distributional impacts of climate policy, attention typically focuses on empirically estimating the effects on the incomes of different population groups, rather than how those groups *perceive* the distributional effects, which is influenced by the narrative and politics of the reform, as outlined in the 4i Framework (see chapter 1). Public opposition can be motivated by a lack of trust and often comes from groups that believe—rightly or wrongly—that a policy will adversely affect them. Even when a policy is designed to benefit them, people may not know it because of poor public communication, or they may not trust the government to deliver effectively. Thus, pairing the analysis of the distributional effects of climate policies with a solid understanding of the public perceptions of the fairness and accountability of government policy making can be crucial for determining how climate policies will be received.

Public perceptions can be important drivers of opposition to climate policies, especially when winners feel like losers, and even when policies follow a sound and progressive design. El Salvador's 2011 gas subsidy reform illustrates this point (Calvo-Gonzales, Cunha, and Trezzi 2015). Although the reform increased the welfare of households in all but the top two deciles of the income distribution, it was unpopular, especially among the lower-income groups who were set to benefit most. Analysis of household surveys indicated that public dissatisfaction before implementation was rooted in misinformation, mistrust in the government's ability to implement the policy, and political beliefs. Perceptions improved gradually—and significantly—over time as households benefited from the reform, with the share of people expressing support for the policy increasing from just 30 percent at the start of implementation to about 65 percent within a year and a half (figure 5.1). This case shows that even beneficiaries of well-designed pro-poor policy reforms may not perceive themselves as winners, so reform strategies need to address information constraints and prior perceptions. Winning public trust can be key.

Empirical evidence from Indonesia shows that opposition to fossil fuel subsidy reform is directly linked to local perceptions of corruption (Kyle 2018). As illustrated in figure 5.2, when corruption levels are perceived to be low, poor households are more than two-and-a-half times more likely to support than to oppose fuel subsidy reform. When corruption is perceived to be high, support declines by 18 percentage points and opposition increases by 14 percentage points. Clearly, public perceptions matter. People's ability to support a subsidy reform requires their confidence that the proposed reform is in their interest and that promised compensation payments will materialize when the subsidy is removed. Governments need to earn public trust: without it, even well-designed, well-intentioned promises of compensation and redistribution can lack credibility.

FIGURE 5.1. Share of population that reported being "satisfied" or "very satisfied" with El Salvador's 2011 subsidy reform, 2011–13

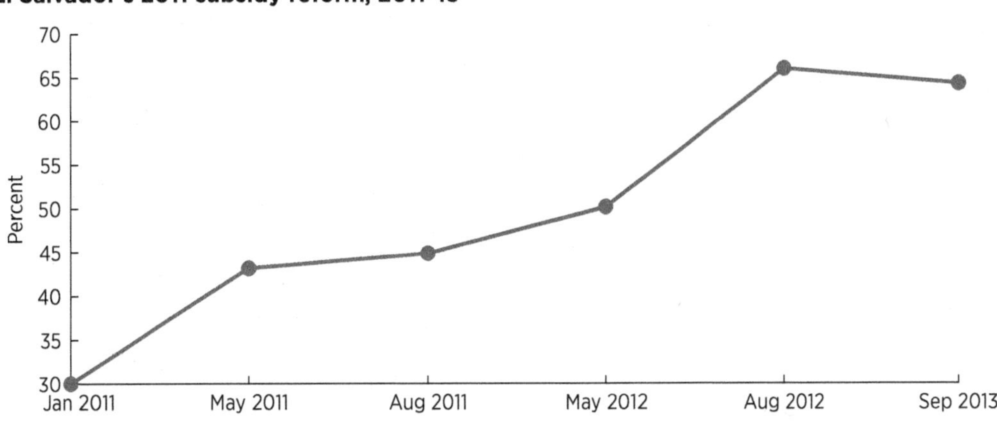

Source: Calvo-Gonzalez, Cunha, and Trezzi 2015.

FIGURE 5.2. **Support for and opposition to subsidy reform in Indonesia, by perceived level of corruption**

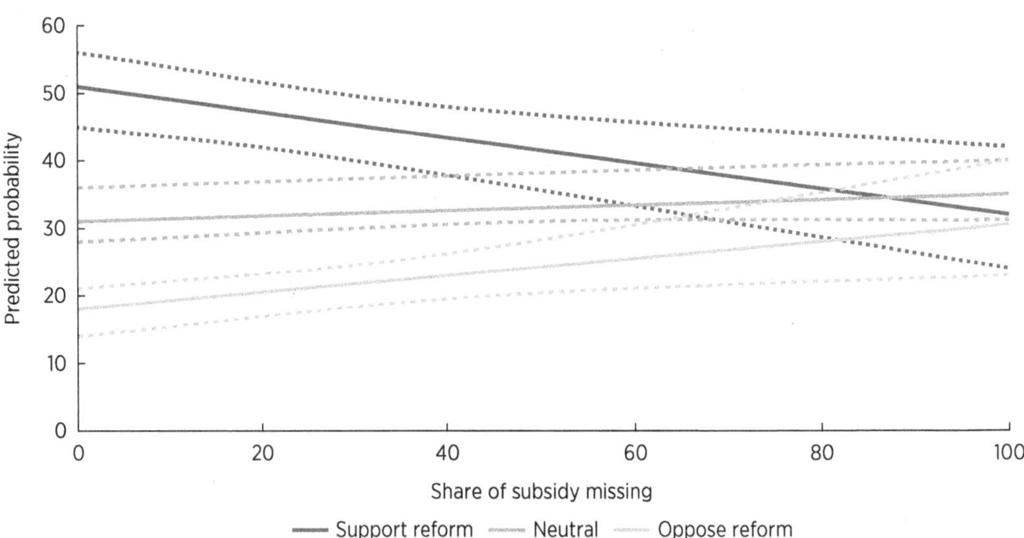

Source: Kyle 2018.
Note: Share of subsidy missing is a proxy for misappropriation of subsidy funds and corruption. A value of 100 suggests the highest level of misappropriation (and 0 the lowest).

Building public trust necessitates understanding and mitigating the impacts of subsidy reform on the poorest and most vulnerable groups, not only to protect livelihoods and ensure a pro-poor reform but also to galvanize support from these groups. Spatial inequalities, social marginalization, and low-income status may cause certain population groups to suffer disproportionately from subsidy reforms. For example, identifying groups that will be harmed by subsidy removal—such as low-income urban taxi drivers, who could experience significant shocks to their disposable income—and devising adequate compensation and social protection schemes are vital. Governments should pay special attention to what stakeholders perceive as negative externalities of the reform. In doing so, governments can address those externalities in a targeted way and mitigate their consequences, thus avoiding situations in which perceptions that compensation is inadequate increase opposition to the reform.

Governments can alleviate credibility concerns by promptly issuing compensation and social protection payments, even before raising carbon taxes or reducing fuel subsidies, as was done in Iran in 2010 (IMF 2013). To convincingly demonstrate their commitment, governments should design such payments to ensure they adequately address the needs of affected stakeholders and should continue to pay them for as long as required to help protect vulnerable livelihoods. Starting complementary and revenue reinvestment measures early will also affirm the government's commitment to the prudent use of reform revenues in the public interest. Carefully considering compensation and using it in conjunction with other measures such as information campaigns and public engagement processes will avoid giving the impression that the government is trying to buy approval.

Transparency in contracts and procurement is essential for gaining people's support for policy reform, it can reduce misconduct, and it is necessary for fair competition. Unlike in oil, mining, and sovereign debt contracts, transparency in energy purchasing contracts is not a globally practiced norm, particularly in emerging markets (Moss and Ibrahim-Tanko 2022). Because many of these contracts involve public money and liabilities, secrecy spurs public distrust, uncertainty, and anger, but also inefficiencies.[2] For example, between 2011 and 2016, the Ghanaian government signed 43 power purchasing agreements for more power than necessary; because some of these contracts included excess capacity charges, the Ministry of Finance paid about US$620 million for unused energy in 2019 (Moss and Ibrahim-Tanko 2022). Even if disclosing such agreements does not fully remove contract risks, if practiced consistently, it can reduce them. This would have a positive effect on people's trust in the system, limit corruption or capture, provide incentives for competition, and accelerate the expansion of reliable and clean energy (Ibrahim-Tanko and Moss 2022).[3]

Using civic engagement to increase support through the policy process

Intergovernmental organizations and scholars have long advocated for civic engagement in climate policy. The 1992 Rio Declaration on Environment and Development has explicit goals for information sharing and citizen participation, the 1997 Kyoto Protocol highlights the importance of public awareness and access to information, and the 2015 Paris Agreement affirms "the importance of education, training, public awareness, public participation, public access to information and cooperation at all levels." In the academic literature, a growing number of publications relate to public engagement and participation in climate action (Hügel and Davies 2020).

But civic engagement can be costly, time-consuming, and resource heavy as such, it comes up against capacity or political resource constraints. This is especially true for climate policy because of its technical complexity, which can be compounded by lower literacy levels, physical mobility, or internet access, and because of the time and effort required to understand problems and participate in decision-making. Involving marginalized and vulnerable groups can be especially difficult where civic engagement efforts have previously failed or disappointed (Wesselink et al. 2011). Successful civic engagement depends on engaging all actors, even those who may have become disengaged or disaffected. Targeting those who are "willing but unable" as well as those who are "able but unwilling" (OECD 2017) may require policy makers to adopt a different set of strategies and approaches.

Civic engagement that is not perceived to be meaningful or fair—for example, because of lack of influence—could result in wasted effort or undesirable outcomes (Gaventa and Barrett 2010; Wamsler et al. 2020). If treated as a mere formality or bureaucratic afterthought, it is unlikely to provide positive outcomes and could risk increasing opposition, especially if it looks like an attempt to deflect protests. A lack of fairness can also be a major issue, for example, when certain groups, such as Indigenous populations or future generations who lack means to participate in traditional engagement processes, are unjustifiably excluded from the process or when citizens' voice is weakened by powerful interest groups. But (un)fairness can also relate to other procedural processes, including rushed proceedings. In the US state of North Carolina, for example, those who attended hearings on new river system pollution control regulations found the process unfair because it did not give them enough time to digest complex technical information (Maguire and Lind 2004).

Civic engagement is especially effective when used throughout the policy-making process, from design through implementation to transition. Box 5.2 presents several examples of public stakeholder engagement at different stages of the decision-making process. The rest of this chapter looks at how civic engagement can help policy makers navigate the political economy across these stages.

Using civic inputs to improve policy design and implementation

Engaging different actors in policy design is a way to ensure that policies and programs account for the priorities and values of—and potential effects on—different groups. For example, in the United States, the District of Columbia held extensive community consultations through multiple conversations with citizens at various locations and events, focus groups, and phone polling when developing its *Sustainable DC 2.0 Plan* (Government of the District of Columbia 2021). Citizens could propose direct edits to the draft plan via an online platform, and the process made efforts to ensure participation of underrepresented groups. This engagement allowed policy makers to tap into the wide range of

BOX 5.2

Deliberative and inclusive policy processes governments can use to engage citizens on climate action

Citizens' juries. These typically involve a small, representative group of lay participants convened to consider a particular question or issue. Over several days, participants receive, cross-question, discuss, and evaluate "evidence" from experts and are then invited to make recommendations. A report is drawn up reflecting their views, including any differences of opinion.

Multicriteria mapping. This methodology combines the transparency and clarity of statistical approaches with the unconstrained framing of open-ended deliberations. After selecting a topic area and defining basic policy options, researchers interview participants individually to develop more policy options and define evaluative criteria, scoring the options and applying relative weightings to the criteria. Participants then come together to discuss the researchers' preliminary quantitative and qualitative analysis, leading to a final report. This approach has been used to explore energy transitions (Chilvers et al. 2021).

Scenario workshops or visioning exercises. These methods allow participants to articulate their vision of the future and consider the kind of future they would like to create. The activities can be applied to broad strategic questions or specific local or sectoral issues. The Transition Network uses such exercises to help articulate what a postcarbon world might look like (Hopkins 2019).

Standing consultative panels or citizens' panels. Normally large representative groups of citizens, these panels are consulted periodically, with a proportion of members replaced at regular intervals. Panels can be used to sample changing opinions and attitudes about a range of issues over time, such as the United Kingdom's standing Peoples' Panel, whose 5,000 randomly selected members of the public are consulted on key issues to track how and why views are changing and to conduct surveys. They can also have a more proactive and policy-facing role, as in Costa Rica's Consejo 5C (see box 5.3).

Ombudsperson for Future Generations. Creating an Ombudsperson for Future Generations is a way to support civic engagement with younger populations. Australia, Hungary, Israel, and Wales are among the countries that have introduced such positions to increase younger generations' influence over government decisions. Unlike traditional ombudspersons, who often focus on maladministration, individual complaints, and the failure of government procedures, an Ombudsperson for Future Generations represents a collection of interests from people not directly represented through democratic process and policy decision-making.

Source: Barron et al. 2023.

on-the-ground experiences and perspectives, and to use them to improve the design of the plan.

Civic engagement can bring different communities' policy concerns to the surface, allowing policy makers to identify where complementary measures are needed. Urban greening policies, for example, are often associated with gentrification (Anguelovski et al. 2022; Derickson, Klein, and Keeler 2021), even in relatively small projects (Anguelovski et al. 2022; Derickson, Klein, and Keeler 2021; Rigolon and Nemeth 2019). In the United States, for example, green development initiatives in Atlanta had significant implications for housing affordability, with housing values increasing up to 26.6 percent (Immergluck and Balan 2017). By engaging affected citizens, policy makers can gain a better understanding of some of the factors contributing to green gentrification and identify measures that can address social inequities in urban green development. Citizen perceptions in gentrified neighborhoods often relate not only to residential and social displacement but also to racial disparities and social cohesion (Bernstein and Isaac 2021). In the face of these additional layers of complexity, civic engagement methods are even more important.

Creating spaces where citizens can learn about and be involved in climate governance decisions can help governments develop acceptable, durable, and effective climate change framework legislation and long-term strategies. Countries are increasingly creating citizens assemblies, platforms, commissions, and other spaces to feed into or produce climate governance frameworks, such as climate laws and strategies (box 5.3). For example, the French Citizens' Assembly was characterized by sustained interactions between citizens and the steering board, with significant input from technical and legal experts, and a strong emphasis on creating consensus. As a result, the citizens adopted 149 recommendations for the government to translate into law (Giraudet et al. 2022). Nevertheless, citizen assemblies like these have often been criticized for granting only limited influence to civic actors, who usually self-select to participate in the process, leaving many lay-citizens who do not proactively engage out of the discussion.

Civic engagement can also help spur policy innovation. Citizen science—that is, scientific research conducted in collaboration with the public—is an effective civic engagement method that can spur innovative solutions to climate challenges. Research and innovation are key to solving the climate crisis, and citizen science offers an effective

BOX 5.3

Consejo 5C: Citizens' Advisory Council on Climate Change

In 2018, the Costa Rican Ministry of Environment and Energy created the Citizens' Advisory Council on Climate Change, also known as Consejo 5C, to fulfill of one of its commitments under the Paris Agreement.[a] With representatives from a cross-section of civil society, Indigenous, business, and trade union groups engaged with climate issues, the council aims to act as a deliberative space that advises the Ministry of Environment and Energy on issues related to climate change. The ministry and other government entities may also submit consultations to the council on the design, implementation, and evaluation of their climate policies, programs, and metrics. The council meets regularly and monitors the country's implementation of its nationally determined contribution and climate change-related Sustainable Development Goal commitments.

a. Information from LATINNO, Citizens' Climate Change Advisory Council, https://latinno.net/en/case/6115/.

approach to leveraging the collective intelligence of citizens (Warin and Delaney 2020). The RISE program's flood monitoring project in Fiji and Indonesia offers an example of a citizen science initiative in which citizens use smartphone technology to gather flood data. Their active contribution to the development of a database documenting water levels and risk zones to support infrastructure planning, and their provision of evidence to validate flood models, also empowers citizens to advocate for infrastructure improvement (Wolff 2021). Such co-creation of knowledge more likely reflects the discourse of diverse groups, leading to more successful policy design than traditional top-down approaches (Coggan et al. 2021).

Community-led development supports local empowerment, capacity building, and efficient resource allocation. Given appropriate technical and financial support, access to information, and clear and transparent rules, communities can efficiently identify priorities, allocate resources, and address development challenges in partnership with local governments and other institutions.[4] Following a community-led, bottom-up approach can better target climate adaptation measures to address local needs. For example, with support from the National Biodiversity Centre of Bhutan, farmers set up community seed banks to maintain buckwheat varieties and enhance the area's genetic diversity by restoring close-to-extinct varieties, increasing the adaptive capacity of local agriculture (Vernooy et al. 2017). In Zambia, unconditional cash transfers in the aftermath of agricultural production or price shocks have empowered rural households to use coping strategies, substantially increasing their food consumption and overall food security (Lawlor et al. 2017).

Giving citizens a greater voice can help governments design climate institutions that resonate with societal concerns, values, and aspirations, but lack of follow-through creates risks. France provides a case in point. After months of debate in Parliament, the law voted on in July 2021 included only about a third to a half of the 146 measures proposed by the Citizen's Assembly (Giraudet et al. 2022). Dismissal of the outcomes and recommendations of civic engagement bodies and processes can undermine climate action going forward and lead to public dissatisfaction. At the same time, however, it is crucial that these bodies do not duplicate or replace the role of existing structure, including parliaments and other decision-making bodies, and that governments carefully and transparently manage expectations about the process for deciding whether to implement recommendations from a specific consultative body.

Communicating rationale, design, and risks to increase public support

Communication helps build and sustain support when implementing climate policies. As outlined by the World Bank's Energy Subsidy Reform Assessment Framework (Worley, Pasquier, and Ezgi 2018), best practice involves integrating a policy communication strategy throughout the planning, development, and implementation stages of a reform; this strategy must also address all affected and interested stakeholders, as well as those who can influence its success. It is important to design communication strategies to reach all segments of society—including people with disabilities, people who are illiterate, or those with limited access to media—and to tailor strategies to the opinions and attitudes of the stakeholders, identified in the realms of opinion research, including, for example, focus group consultations and public opinion surveys (see Worley, Pasquier, and Ezgi 2018).

Despite the usefulness of civic engagement methods in addressing concerns on a local level, increasing public support for policy reform at a larger scale requires strategic communication. Participatory approaches to civic engagement—such as consensus building in Ghana (Centre for Public Impact and Calouste Gulbenkian Foundation 2021), participatory green budgeting in Guinea (Oshima and Perrin 2018), community-owned renewable energy projects in the United Kingdom,[5] and participatory wind turbine siting negotiations in the United States (Firestone et al. 2020)—promote a sense of ownership or agency, thus improving implementation outcomes. Such approaches tend to focus on small-scale local projects and are less well geared to increase support for broader, national-level climate policies such as carbon pricing reforms or energy efficiency regulations. But governments have ways, including communication strategies, to reinstate agency, particularly if those strategies are complemented by behavioral interventions. Energy labels, feedback devices, and nudging, for example, are popular policy measures used to reduce energy demand (Composto and Weber 2022; Cornago 2021).

Clear communication is essential for a successful civic engagement process. Because civic engagement interventions range from empowering participants in decision-making to raising awareness and providing information, the objectives of engagement can vary and are not always clear. To manage expectations and make the most of civic engagement, clearly communicating the design and objectives of the process is vital to ensure participants know exactly what is expected of them and how their engagement will inform policy making, facilitating a process that is unambiguous, transparent, and effective (Uittenbroek et al. 2019). This communication includes explaining how stakeholder and civic society inputs will be or have been considered in decision-making and providing explanations when the outcome does not reflect those inputs (Lind and Arndt 2016). For example, the Colombian Environment Ministry publishes responses to comments online and indicates, with an explanation, whether a comment is accepted or rejected; in Costa Rica, responses are sent to individuals via email and comments and responses are made available online. At the same time, it is important to establish venues that allow citizens to identify policy issues that are a priority for them. Such venues can prevent civic engagement on specific issues from being tainted by other topics that citizens find more pressing.

When well designed, communication can help policy makers navigate the political economy and build support for key reforms. For example, in the lead-up to and after winning national elections in 2014, the Indonesian president communicated the regressive nature of subsidies to the electorate and offered targeted support to other areas, such as education and health care, to build support for the reforms; the government implemented those reforms when oil prices were low and negative effects would be limited. And, before removing most of its fossil fuel subsidies in 2015 (under International Monetary Fund conditions), Ghana carried out extensive stakeholder engagement and communication campaigns that explained the need for reform and collected inputs for reform design, including an exemption for low-octane fuel used by politically important coastal fishing communities (McCulloch 2023).

Developing comprehensive communication strategies and building an institutional support structure can help counteract misinformation. Reaching skeptical audiences can be difficult, because individuals often seek information from sources that reinforce their worldviews and are subject to confirmation bias effects (Newman, Nisbet, and Nisbet 2018). This tendency provides fertile ground for lobbies that direct considerable resources to influencing public perceptions and opinions, including through misinformation campaigns (Farrell 2016; Lewandowsky 2021; Moreno, Kinn, and Narberhaus 2022). For example, research has revealed that 80 percent of Exxon Mobil's internal documents between 1977 and 2014 acknowledged human-caused climate change but 81 percent of its public-facing materials expressed doubt, and that the company directed significant funding to think tanks known for producing research that misrepresented the science on climate change (Oreskes and Conway 2010; Supran and Oreskes 2017). Useful strategies for protecting populations against misinformation include emphasizing scientific consensus, highlighting risks of misinformation and preemptive refutation (van der Linden et al. 2017), having an authority figure trusted by skeptics correcting misinformation (Benegal and Scruggs 2018), limiting misinformation networks—for example, using litigation to hold vested interest lobby groups to account when spreading misinformation—and increasing transparency in lobbying (Farrell, McConnell, and Brulle 2019).

Climate policy choices and communication should consider actors' motivation for pro-climate actions. People, especially younger generations, are increasingly adopting plant-based diets, using low-carbon modes of transportation, recycling, and purchasing green label products and secondhand clothes. Motivations for adopting green lifestyle choices can be *intrinsic* (personally rewarding) and *extrinsic* (for example, in response to an external financial incentive). But it is generally acknowledged that intrinsic motivation is the leading force behind pro-environmental behavior (Silvi and Padilla 2021) and that extrinsic motivation induces only short-term behavior change (van der Linden 2015). The behavioral economics literature also suggests that extrinsic motivation can crowd out intrinsic motivation for pro-environmental behavior (Rode, Gomez-Baggethun, and Krause 2015). For example, a randomized controlled trial in Ecuador found that households that received a letter comparing their energy use to that of the average household (intrinsic motivation) consumed less energy in the postintervention period, whereas those who received a letter that also had information on expected energy savings (extrinsic motivation) did not change their consumption behavior (Pellerano et al. 2017). Market-based instruments, such as cap-and-trade mechanisms or environmental taxes, are also prone to crowding out (Cinner et al. 2020).

Civic engagement can help prevent crowding out and can crowd in more pro-climate action among citizens. Crowding out can result from several psychological mechanisms, including reduced internal satisfaction, reduced sense of agency, and reduced moral responsibility. Adopting multilevel decision-making approaches that involve public, private, and community actors can help reduce the crowding-out effect (Ostrom 2002). Different forms of citizen engagement can help restore intrinsic motivation. For example, engagement that enhances the level of information that citizens have about the behavior of others can increase willingness to contribute to public goods, including climate change (Fischbacher, Gaechter, and Fehr 2001; Schleich, Schwirplies, and Ziegler 2017). Maintaining intrinsic motivation is important to ensure successful implementation of climate policies. On the flip side, individual- or household-level actions guided by intrinsic motivation to climate action risk crowding out support for national-level policy (Knook, Dorner, and Stahlmann-Brown 2022). Information and awareness about the relative potential of different measures, and the importance of national-level interventions, can help.

Information sharing and awareness raising can help shift preferences and shape behavior. For example, the government of Chile met initial resistance from Indigenous communities to developing geothermal energy projects in their territories. But communication and engagement efforts, focused on raising awareness, showed that the resistance stemmed from misperceptions about the impacts and potential benefits of the development (World Bank 2021). After these misperceptions were addressed through workshops, geothermal plant tours, and other initiatives, local support increased significantly. Engaging effectively with citizens on national-level policies can be more challenging because of those policies' inherent complexities and extended scales, but methods that rely on awareness raising and education can help encourage and empower citizens to have more informed discussions (Pidgeon et al. 2014). The United Kingdom's Climate Assembly, for example, brought together citizens to discuss how the country can meet its net zero emissions target. Participants received comprehensive information on the different ways to achieve net zero, enabling them to make informed recommendations on complex policy issues (Cherry et al. 2021, Climate Assembly UK, n.d.). Nevertheless, such engagement is no silver bullet. For example, the steady increase of the carbon tax in France has been met with suspicion by many citizens—despite public engagement and information sharing—because they perceived that it was implemented to meet the needs of the government's general budget rather than to reduce greenhouse gas emissions (Bureau, Henriet, and Schubert 2019). In cases where awareness raising is insufficient, support can also be increased by ensuring that complementary measures to alleviate distributional implications are in place and that these measures are clearly communicated to those affected by the policy in question.

Tackling trade-offs and risks is part of effective communication and can build trust and support. Research shows that communicating not only how a public policy works but also its associated risks and trade-offs can help increase policy support (OECD 2021)—and climate policies are no exception. An experimental study across 20 countries finds that providing information on the effectiveness and distributional implications of climate policies significantly improves support, whereas providing people with information on the implications of climate change has little or no significant impact (Dechezleprêtre et al. 2022).

Notes

1. The NIMBY—from "not in my backyard"—concept is used to explain public opposition to new developments considered undesirable near people's homes and communities (Devine-Wright 2009).
2. From Energy for Growth Hub's "Contract Transparency" web page, https://energyforgrowth.org/project/contract-transparency/.
3. Energy for Growth Hub, "Contract Transparency."
4. From the World Bank's "Community and Local Development" web page, https://www.worldbank.org/en/topic/communitydrivendevelopment#1.
5. See the UK Government's "Community Energy" web page (last updated January 26, 2015), https://www.gov.uk/guidance/community-energy.

References

Anguelovski, I., J. J. T. Connolly, H. Cole, M. Garcia-Lamarca, M. Triguero-Mas, F. Baro, N. Martin, et al. 2022. "Green Gentrification in European and North American Cities." *Nature Communications* 13 (3816). https://doi.org/10.1038/s41467-022-31572-1.

Armeni, C., and M. Lee. 2021. "Participation in a Time of Climate Crisis." *Journal of Law and Society* 48 (4): 549–72. https://doi.org/10.1111/jols.12320.

Barron, P., L. Cord, J. Cuesta, S. A. Espinoza, G. Larson, and M. Woolcock. 2023. *Social Sustainability in Development: Meeting the Challenges of the 21st Century.* Washington, DC: World Bank. https://doi.org/10.1596/978-1-4648-1946-9.

Benegal, S. D., and L. A. Scruggs. 2018. "Correcting Misinformation about Climate Change: The Impact of Partisanship in an Experimental Setting." *Climatic Change* 148 (1): 61–80. https://doi.org/10.1007/s10584-018-2192-4.

Bernstein, A. G., and C. A. Isaac. 2021. "Gentrification: The Role of Dialogue in Community Engagement and Social Cohesion." *Journal of Urban Affairs* 45 (4): 753–70. https://doi.org/10.1080/07352166.2021.1877550.

Bureau, D., F. Henriet, and K. Schubert. 2019. "Pour le climat: une taxe juste, pas juste une taxe." *Notes du conseil d'analyse économique* 50: 1–2. https://doi.org/10.3917/ncae.050.0001.

Calvo-Gonzales, O., B. Cunha, and R. Trezzi. 2015. "When Winners Feel Like Losers: Evidence from an Energy Subsidy Reform." Policy Research Working Paper 7265, World Bank, Washington, DC. http://hdl.handle.net/10986/21998.

CAST (Centre for Climate Change and Social Transformations). 2021. "Public Perceptions of Climate Change and Policy Action in the UK, China, Sweden and Brazil." CAST briefing 10. https://cast.ac.uk/wp-content/uploads/2021/10/01112021-Briefing-10-final.pdf.

Centre for Public Impact and Calouste Gulbenkian Foundation. 2021. *Public Engagement on Climate Change. A Case Study Compendium.* Centre for Public Impact. https://www.centreforpublicimpact.org/assets/documents/cpi-cgf-public-engagement-climate-change-case-studies.pdf.

Cherry, C., S. Capstick, C. Demski, C. Mellier, L. Stone, and C. Verfuerth. 2021. *Citizens' Climate Assemblies: Understanding Public Deliberation for Climate Policy*. Monograph. Cardiff: Cardiff University. https://orca.cardiff.ac.uk/id/eprint/145771/.

Chilvers, J., R. Bellamy, H. Pallett, and T. Hargreaves. 2021. "A Systemic Approach to Mapping Participation with Low-Carbon Energy Transitions." *Nature Energy* 6 (3): 250–59. https://doi.org/10.1038/s41560-020-00762-w.

Cinner, J. E., M. L. Barnes, G. G. Gurney, S. Lockie, and C. Rojas. 2020. "Markets and the Crowding Out of Conservation-Relevant Behavior." *Conservation Biology* 35 (3): 816–23. https://doi.org/10.1111/cobi.13606.

Climate Assembly UK. n.d. *The Path to Net Zero.* https://www.climateassembly.uk/report/read/final-report.pdf.

Coggan, A., J. Carwardine, S. Fielke, and S. Whitten. 2021. "Co-creating Knowledge in Environmental Policy Development. An Analysis of Knowledge Co-Creation in the Review of the Significant Residual Impact Guidelines for Environmental Offsets in Queensland, Australia." *Environmental Challenges* 4: 100138. https://doi.org/10.1016/j.envc.2021.100138.

Composto, J. W., and E. U. Weber. 2022. "Effectiveness of Behavioural Interventions to Reduce Household Energy Demand: A Scoping Review." *Environmental Research Letters* 17 (6). https://doi.org/10.1088/1748-9326/ac71b8.

Cornago, E. 2021. "The Potential of Behavioural Interventions for Optimising Energy Use at Home." International Energy Agency, Paris. https://www.iea.org/articles/the-potential-of-behavioural-interventions-for-optimising-energy-use-at-home.

Dechezleprêtre, A., A. Fabre, T. Kruse, P. Bluebery, A. Sanchez Chico, and S. Stantcheva. 2022. "Fighting Climate Change: International Attitudes Toward Climate Policies." OECD Economics Department Working Papers 1714, OECD Publishing, Paris.

Derickson, K., M. Klein, and B. L. Keeler. 2021. "Reflections on Crafting a Policy Toolkit for Equitable Green Infrastructure." *npj Urban Sustainability* 1 (21). https://doi.org/10.1038/s42949-021-00014-0.

Devine-Wright, P. 2009. "Rethinking NIMBYism: The Role of Place Attachment and Place Identity in Explaining Place-Protective Action." *Journal of Community and Applied Social Psychology* 19 (6): 426–41. https://doi.org/10.1002/casp.1004.

Doberstein, C. 2020. "Role-Playing in Public Engagement for Housing for Vulnerable Populations: An Experiment Exploring its Possibilities and Limitations." *Land Use Policy* 99: 105032. https://doi.org/10.1016/j.landusepol.2020.105032.

Farrell, J. 2016. "Network Structure and Influence of the Climate Change Counter-Movement." *Nature Climate Change* 6 (4): 370–74. https://doi.org/10.1038/nclimate2875.

Farrell, J., K. McConnell, and R. Brulle. 2019. "Evidence-Based Strategies to Combat Scientific Misinformation." *Nature Climate Change* 9 (3): 191–95. https://doi.org/10.1038/s41558-018-0368-6.

Firestone, J., C. Hirt, D. Bidwell, M. Gardner, and J. Dwyer. 2020. "Faring Well in Offshore Wind Power Siting? Trust, Engagement and Process Fairness in the United States." *Energy Research and Social Science* 62: 101393. https://doi.org/10.1016/j.erss.2019.101393.

Fischbacher, U., S. Gaechter, and E. Fehr. 2001. "Are People Conditionally Cooperative? Evidence from a Public Goods Experiment." *Economics Letters* 71 (3): 397–404. https://doi.org/10.1016/S0165-1765(01)00394-9.

Fischer, T. B. 2022. "Replacing EIA and SEA with Environmental Outcome Reports (EORs)—The 2022 Levelling Up and Regeneration Bill in the UK." *Impact Assessment and Project Appraisal* 40 (4): 267–68. https://doi.org/10.1080/14615517.2022.2089375.

Gaventa, J., and G. Barrett. 2010. "So What Difference Does It Make? Mapping the Outcomes of Citizen Engagement." Working Paper 347, Institute of Development Studies, University of Sussex, Brighton, U.K. https://www.ids.ac.uk/download.php?file=files/dmfile/Wp347.pdf.

Geißler, G., and A. Jiricka-Pürrer. 2023. "The Future of Impact Assessment in Austria and Germany – Streamlining Impact Assessment to Save the Planet?" *Impact Assessment and Project Appraisal* 41 (3): 215–22. https://doi.org/10.1080/14615517.2023.2186595.

Giraudet, L.-G, B. Apouey, H. Arab, S. Baeckelandt, P. Bégout, N. Berghmans, N. Blanc, et al. 2022. "'Co-Construction' in Deliberative Democracy: Lessons from the French Citizens' Convention for Climate." *Humanities and Social Sciences Communications* 9 (1): 1–16. https://doi.org/10.1057/s41599-022-01212-6.

Gonzales, A., and I. Sobrini. 2023. "Environmental Assessment Simplification in Spain: Streamlining or Weakening Procedures?" *Impact Assessment and Project Appraisal* 41 (3): 190–93. https://doi.org/10.1080/14615517.2023.2170094.

Government of the District of Columbia. 2021. *Sustainable DC 2.0 Plan*. https://sustainable.dc.gov/sdc2.

Hopkins, R. 2019. *From What Is to What If: Unleashing the Power of Imagination to Create the Future We Want*. Chelsea Green Publishing.

Hügel, S., and A. R. Davies. 2020. "Public Participation, Engagement and Climate Change Adaptation: A Review of the Research Literature." *WIREs Climate Change* 11 (4). https://doi.org/10.1002/wcc.645.

Ibrahim-Tanko, R., and T. Moss. 2022. "How Power Contract Transparency Can Help Fight Poverty and Climate Change." *Forbes*, June 1, 2022. https://www.forbes.com/sites/thebakersinstitute/2022/06/01/how-power-contract-transparency-can-help-fight-poverty-and-climate-change/?sh=94c1951156de.

Immergluck, D., and T. Balan. 2017. "Sustainable for Whom? Green Urban Development, Environmental Gentrification and the Atlanta Beltline." *Urban Geography* 39 (4): 546–62. https://doi.org/10.1080/02723638.2017.1360041.

IMF (International Monetary Fund). 2013. Energy Subsidy Reform: Lessons and Implications. Washington, DC: IMF.

ITS (Instituto de Tecnologia e Sociedade do Rio. 2021 . "Brazilian Citizen's Perception towards Climate Change." Study. ITS, Rio de Janeiro. https://itsrio.org/wp-content/uploads/2022/03/Climate-Change-and-Public-Perception-in-Brazil_2021.pdf.

Knook, J., Z. Dorner, and P. Stahlmann-Brown. 2022. "Priming for Individual Energy Efficiency Action Crowds Out Support for National Climate Change Policy." *Ecological Economics* 191:107239. https://doi.org/10.1016/j.ecolecon.2021.107239.

Kyle, J. 2018. "Local Corruption and Popular Support for Fuel Subsidy Reform in Indonesia." *Comparative Political Studies* 51 (11). https://doi.org/10.1177/0010414018758755.

Lamb, W. F., and J. C. Minx. 2020. "The Political Economy of National Climate Policy: Architectures of Constraint and a Typology of Countries." *Energy Research and Social Science* 64: 101429. https://doi.org/10.1016/j.erss.2020.101429.

Lawlor, K., S. Handa, D. Seidenfeld, and The Zambia Cash Transfer Evaluation Team. 2017. "Cash Transfers Enable Households to Cope with Agricultural Production and Price Shocks: Evidence from Zambia." *Journal of Development Studies* 55 (2): 209–26. https://doi.org/10.1080/00220388.2017.1393519.

Lewandowsky, S. 2021. "Climate Change Disinformation and How to Combat It." *Annual Review of Public Health* 42 (1): 1–21. https://doi.org/10.1146/annurev-publhealth-090419-102409.

Lind, E. A., and C. Arndt. 2016. "Perceived Fairness and Regulatory Policy: A Behavioural Science Perspective on Government-Citizen Interactions." OECD Regulatory Policy Working Paper No. 6, OECD Publishing, Paris. https://dx.doi.org/10.1787/1629d397-en.

Maguire, L. A., and E. A. Lind. 2004. "Public Participation in Environmental Decisions: Stakeholders, Authorities and Procedural Justice." *International Journal of Global Environmental Issues* 3 (2): 133–48. https://doi.org/10.1504/IJGENVI.2003.003861.

McCulloch, N. 2023. *Ending Fossil Fuel Subsidies: The Politics of Saving the Planet.* Practical Action Publishing.

Moreno, J. A., M. Kinn, and M. Narberhaus. 2022. "A Stronghold of Climate Change Denialism in Germany: Case Study of the Output and Press Representation of the Think Tank EIKE." *International Journal of Communication* 16: 22.

Moss, T., and R. Ibrahim-Tanko. 2022. "The Other Hidden Debt—How Power Contract Transparency Can Help Prevent Future Debt Risk." *IMF PFM Blog*, June 21, 2022. https://blog-pfm.imf.org/en/pfmblog/2022/06/the-other-hidden-debt-how-power-contract-transparency-can-help-prevent-future-de.

Newman, T. P., E. C. Nisbet, and M. C. Nisbet. 2018. "Climate Change, Cultural Cognition and Media Effects: Worldviews Drive News Selectivity, Biased Processing and Polarized Attitudes." *Public Understanding of Science* 27 (8): 985–1002. https://doi.org/10.1177/0963662518801170.

OECD (Organisation for Economic Co-operation and Development). 2017. *Preventing Policy Capture. Integrity in Public Decision Making.* Paris: OECD Publishing. https://doi.org/10.1787/9789264065239-en.

OECD (Organisation for Economic Co-operation and Development). 2021. *OECD Regulatory Policy Outlook 2021.* Paris: OECD Publishing. https://doi.org/10.1787/38b0fdb1-en.

Oreskes, N., and E. M. Conway. 2010. *Merchants of Doubt: How a Handful of Scientists Obscured the Truth on Issues from Tobacco Smoke to Global Warming.* Bloomsbury Press.

Oshima, K., and N. Perrin. 2018. "Citizen Engagement in Rural Guinea: Making Tangible Changes from the Bottom Up." *World Bank Blogs*, January 18, 2018. https://blogs.worldbank.org/nasikiliza/citizen-engagement-in-rural-guinea-making-tangible-changes-from-the-bottom-up.

Ostrom, E. 2002. "Crowding Out Citizenship." *Scandinavian Political Studies* 23 (1): 3–16. https://doi.org/10.1111/1467-9477.00028.

Pellerano, J. A., M. K. Price, S. L. Puller, and G. Sanchez. 2017. "Do Extrinsic Incentives Undermine Social Norms? Evidence from a Field Experiment in Energy Conservation." *Environmental and Resource Economics* 67: 413–28. https://doi.org/10.1007/s10640-016-0094-3.

Pidgeon, N., C. Demski, C. Butler, and A. Spence. 2014. "Creating a National Citizen Engagement Process for Energy Policy." *PNAS* 111 (4). https://doi.org/10.1073/pnas.1317512111.

Rigolon, A., and J. Nemeth. 2019. "Green Gentrification or 'Just Green Enough': Do Park Location, Size and Function Affect Whether a Place Gentrifies or Not?" *Urban Studies* 57 (2): 402–20. https://doi.org/10.1177/0042098019849380.

Rode, J., E. Gomez-Baggethun, and T. Krause. 2015. "Motivation Crowding by Economic Incentives in Conservation Policy: A Review of the Empirical Evidence." *Ecological Economics* 117: 270–82. https://doi.org/10.1016/j.ecolecon.2014.11.019.

Rowe, G., T. Horlick-Jones, N. F. Pidgeon, J. Walls, and W. Poortinga. 2008. "Analysis of a Normative Framework for Evaluating Public Engagement Exercises: Reliability, Validity and Limitations." *Public Understanding of Science* 17 (4). https://doi.org/10.1177/0963662506075351.

Schleich, J., C. Schwirplies, and A. Ziegler. 2017. "Do Perceptions of international Climate Policy Stimulate or Discourage Voluntary Climate Protection Activities? A Study of German and US Households." *Climate Policy* 18 (5): 568–80. https://doi.org/10.1080/14693062.2017.1409189.

Silvi, M., and E. Padilla. 2021. "Pro-Environmental Behavior: Social Norms, Intrinsic Motivation and External Conditions." *Environmental Policy and Governance* 31 (6): 619–32. https://doi.org/10.1002/eet.1960.

Supran, G., and N. Oreskes. 2017. "Assessing ExxonMobil's Climate Change Communications (1977–2014)." *Environmental Research Letters* 12 (8): 084019. https://doi.org/10.1088/1748-9326/aa815f.

Uittenbroek, C. J., H. L. P. Mees, D. L. T. Hegger, and P. P. J. Driessen. 2019. "The Design of Public Participation: Who Participates, When and How? Insights in Climate Adaptation Planning from the Netherlands." *Journal of Environmental Planning and Management* 62 (14): 2529–47. https://doi.org/10.1080/09640568.2019.1569503.

UNDP (United Nations Development Programme) and University of Oxford. 2021. *The Peoples' Climate Vote.* UNDP and University of Oxford. https://www.undp.org/publications/peoples-climate-vote.

van der Linden, S. 2015. "Intrinsic Motivation and Pro-Environmental Behaviour." *Nature Climate Change* 5: 612–13. https://doi.org/10.1038/nclimate2669.

van der Linden, S., A. Leiserowitz, S. Rosenthal, and E. Maibach. 2017. "Inoculating the Public against Misinformation about Climate Change." *Global Challenges* 1 (2): 1600008. https://doi.org/10.1002/gch2.201600008.

Vernooy, R., B. Sthapit, G. Otieno, P. Shrestha, and A. Gupta. 2017. "The Roles of Community Seed Banks in Climate Change Adaption." *Development in Practice* 27 (3): 316–27. https://doi.org/10.1080/09614524.2017.1294653.

Wamsler, C., N. Schäpke, C. Fraude, D. Stasiak, T. Bruhn, M. Lawrence, H. Schroeder, and L. Mundaca. 2020. "Enabling New Mindsets and Transformative Skills for Negotiating and Activating Climate Action: Lessons from UNFCCC Conferences of the Parties." *Environmental Science and Policy* 112: 227–35. https://doi.org/10.1016/j.envsci.2020.06.005.

Warin, C., and N. Delaney. 2020. *Citizen Science and Citizen Engagement. Achievements in Horizon 2020 and Recommendations on the Way Forward.* European Commission, Directorate-General for Research and Innovation. https://data.europa.eu/doi/10.2777/05286.

Wesselink, A., J. Paavola, O. Fritsch, and O. Renn. 2011. "Rationales for Public Participation in Environmental Policy and Governance: Practitioners' Perspectives." *Environment and Planning A: Economy and Space* 43 (11). https://doi.org/10.1068/a44161.

Wilson, C. 2018. "Disruptive Low-Carbon Innovations." *Energy Research and Social Science* 37: 216–23. https://doi.org/10.1016/j.erss.2017.10.053.

Wolff, E. 2021. "The Promise of a 'People-Centred' Approach to Floods: Types of Participation in the Global Literature of Citizen Science and Community-Based Flood Risk Reduction in the Context of the Sendai Framework." *Progress in Disaster Science* 10. https://doi.org/10.1016/j.pdisas.2021.100171.

World Bank. 2021. "Citizen Engagement Proves Vital to the Success of Renewable Power Projects in Chile." News release, September 28, 2021. https://www.worldbank.org/en/news/feature/2021/09/28/la-participaci-n-ciudadana-resulta-vital-para-el-xito-de-los-proyectos-de-energ-a-renovable-en-chile.

World Bank. 2023. *Reality Check: Lessons from 25 Policies Advancing a Low-Carbon Future.* Climate Change and Development Series. Washington, DC: World Bank. http://hdl.handle.net/10986/40262.

Worley, H. B., S. M. B. Pasquier, and C. Ezgi. 2018. *Designing Communication Campaigns for Energy Subsidy Reform: Communication.* Energy Subsidy Reform Assessment Framework (ESRAF) Good Practice Note. Washington, DC: World Bank. http://documents.worldbank.org/curated/en/939551530880505644/Designing-Communication-Campaigns-for-Energy-Subsidy-Reform-Communication.

ECO-AUDIT

Environmental Benefits Statement

The World Bank Group is committed to reducing its environmental footprint. In support of this commitment, we leverage electronic publishing options and print-on-demand technology, which is located in regional hubs worldwide. Together, these initiatives enable print runs to be lowered and shipping distances decreased, resulting in reduced paper consumption, chemical use, greenhouse gas emissions, and waste.

We follow the recommended standards for paper use set by the Green Press Initiative. The majority of our books are printed on Forest Stewardship Council (FSC)–certified paper, with nearly all containing 50–100 percent recycled content. The recycled fiber in our book paper is either unbleached or bleached using totally chlorine-free (TCF), processed chlorine–free (PCF), or enhanced elemental chlorine–free (EECF) processes.

More information about the Bank's environmental philosophy can be found at http://www.worldbank.org/corporateresponsibility.